Where
the Birds
Never
Sing

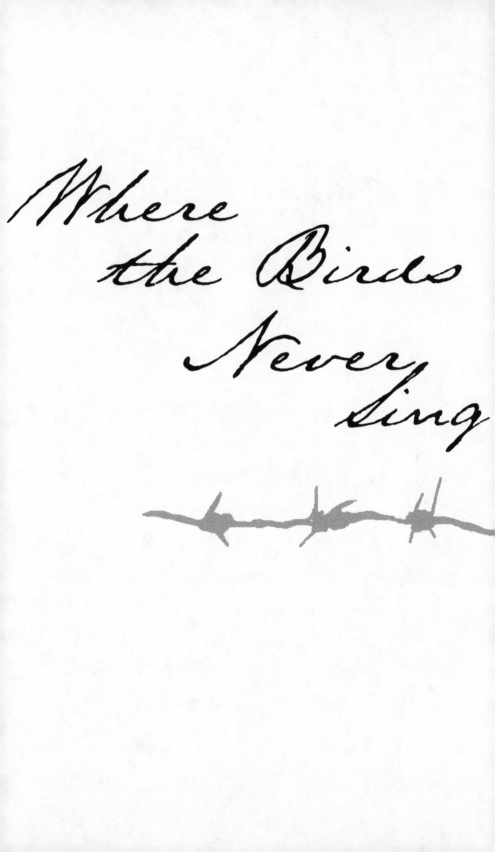

Where
the Birds
Never
Sing

The
True Story
of the
92nd Signal
Battalion
and
the Liberation
of Dachau

Jack Sacco

10 ReganBooks
Celebrating Ten Bestselling Years
An Imprint of HarperCollins*Publishers*

A hardcover edition of this book was published in 2003 by ReganBooks, an imprint of HarperCollins Publishers.

All photographs courtesy of Joe Sacco except for the following: Mrs. Ernest Thomas: page 40; Jim Hodges: pages 94, 135, 173, 176, 249, 257, and 288; Ed Duthie: page 258; Paul Averitt: pages 276 and 277 (both photos); Wilbur Fulton: pages 278 and 287; Joe Hutter: page 306.

HarperCollins books may be purchased for educational, business, or sales promotional use. For information please write: Special Markets Department, HarperCollins Publishers Inc., 10 East 53rd Street, New York, NY 10022.

First paperback edition published 2004

Designed by Nancy Singer Olaguera

Map designed by David Cain

Printed on acid-free paper

The Library of Congress has cataloged the hardcover edition as follows:

Sacco, Jack.
    Where the birds never sing : the true story of the 92nd Signal Battalion and the liberation of Dachau / Jack Sacco.
    p. cm.
ISBN 0-06-009665-9
    1. Sacco, Joe, 1924– 2. United States. Army. Signal Battalion, 92nd. 3. Dachau (Concentration camp) 4. World War, 1939–1945—Regimental histories—United States. 5. World War, 1939–1945—Campaigns—Western Front. 6. Soldiers—United States—Biography.
I. Title.
D769.36392nd. S33 2003
904.54'2133—dc21   2003046646

ISBN 0-06-009666-7 (pbk.)

04 05 06 07 08 QTC/RRD 10 9 8 7 6 5 4 3 2 1

To Joe Sacco and the men of the 92nd Signal Battalion

They traveled far from the homes and families they loved.
They fought and died for people they did not know.
And, in their own simple and profoundly selfless way,
they not only served their country, they saved the world.

They say that the birds never sing at Dachau. Perhaps they cannot produce their wondrous music in a place that has witnessed such tragedy, such cruelty, such horror. Perhaps God forbids it. Or perhaps, on their own, they are muted by the profound sense of sadness that permeates the very air around Dachau—air that once was filled with the cries of innocents and the lingering smoke of their ashes.

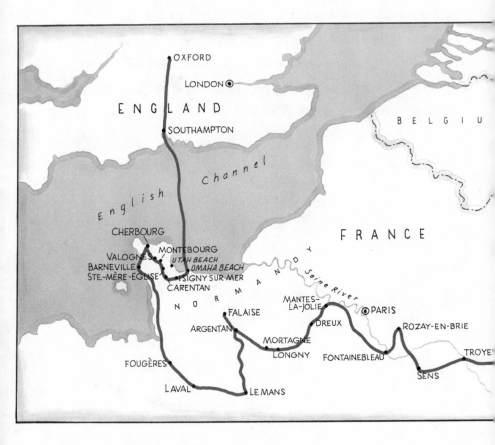

OXFORD

LONDON ◉

E N G L A N D

BELGIU

SOUTHAMPTON

Channel

English

F R A N C E

CHERBOURG

MONTEBOURG
VALOGNES
UTAH BEACH
BARNEVILLE
OMAHA BEACH
STE.-MÈRE-ÉGLISE
ISIGNY SUR MER
CARENTAN

N O R M A N D Y

Seine River

MANTES-
LA-JOLIE
◉ PARIS

FALAISE

ARGENTAN
DREUX
ROZAY-EN-BRIE

MORTAGNE
LONGNY
FONTAINEBLEAU
TROYE

FOUGÈRES
SENS

LAVAL
LE MANS

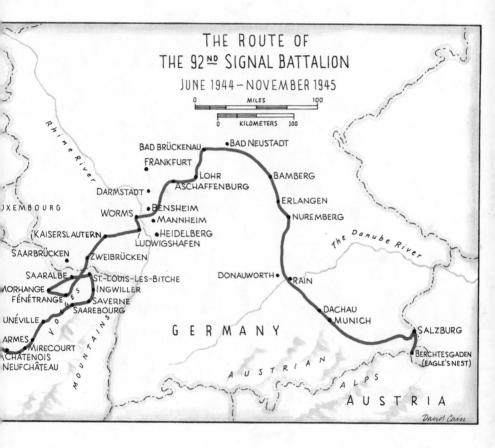

THE ROUTE OF
THE 92ND SIGNAL BATTALION
JUNE 1944 – NOVEMBER 1945

MILES
0   100

KILOMETERS
0   100

Rhine River

• BAD NEUSTADT
BAD BRÜCKENAU •
• FRANKFURT
LOHR •   • BAMBERG
DARMSTADT •   • ASCHAFFENBURG
                              • ERLANGEN
UXEMBOURG   • BENSHEIM   • NUREMBERG
WORMS •   • MANNHEIM
KAISERSLAUTERN • HEIDELBERG
                • LUDWIGSHAFEN
SAARBRÜCKEN
        • ZWEIBRÜCKEN
SAARALBE •   • DONAUWORTH   • RAIN
     • ST.-LOUIS-LES-BITCHE
MORHANGE •   • INGWILLER
FÉNÉTRANGE •
        • SAVERNE
        • SAAREBOURG   • DACHAU
UNÉVILLE •   • MUNICH

GERMANY   • SALZBURG

ARMES •
  • MIRECOURT   • BERCHTESGADEN
CHÂTENOIS   (EAGLE'S NEST)
NEUFCHÂTEAU •

The Danube River

AUSTRIAN   ALPS

AUSTRIA

VOSGES MOUNTAINS

David Cain

Contents

Foreword

In my life, I have been blessed beyond measure with a loving family and staunch friends. For more than thirty-five years the people of my home state entrusted me with the privilege of representing them in Washington. Twice my party has nominated me for the highest office in the land. Yet, no honor that has been bestowed upon me has engendered greater pride than that of wearing my uniform in the service of my country.

When America was called upon to rise up against the tyranny and the almost unthinkable atrocities of the Nazis and their allies, an entire generation responded. Ordinary men, like Jack Sacco's father, Joe, myself, and thousands more rallied to the cause and found ourselves facing hardships the likes of which we had never imagined—and we won.

Some sixty years later, we have been called the Greatest Generation, but our ranks are slowly thining. But just as our cause was worth fighting for, so too are our stories worth remembering. Jack Sacco has paid his father and all those who have ever served this country the ultimate tribute. *Where the Birds Never Sing* is an honest, moving portrait of his father, an American soldier in World War II, who fought for everything that makes our nation great.

This book does not glorify war—far from it. For no one knows better than a soldier that war is in many ways the worst of man's creations. Yet there are principles worth defending and evils that must be stopped. In doing so, the greatest qualities of which human beings are capable emerge: courage beyond measure, loyalty beyond words, sacrifice, ingenuity, endurance, and a love of country and fellow man that is truly boundless.

Thanks to men like Jack Sacco, Americans will never lack for stories of heroes. And thanks to men like his father, Joe, we'll never forget how heroes are made.

—Senator Bob Dole

Introduction

When I was a boy, my father often told me stories about the war. I would listen with wide-eyed fascination as he recounted tales of how he and his buddies fought their way across Europe under the leadership of General George S. Patton. He showed me Nazi swords, daggers, and other artifacts he had collected as his battalion stormed through France and Germany. But there was more to the story than he could share with such a young boy.

One day, just after I turned twelve, he called me into the family den and asked me to sit with him and my mother. He pulled out a small photo album. "I'm going to show you something that happened during the war," he said. He glanced at my mother, who cautiously nodded that it was okay for him to open the book. "I didn't want you to see these until you were old enough," he continued. "My buddies and I took them the day we liberated the concentration camp at Dachau."

"Concentration camp?" I asked.

"Yeah," he answered. "The Nazis killed people there. But we made them stop."

He opened the album and handed it to me. Inside I saw horrific images of suffering, cruelty, and death, the likes of which I could have never imagined. I stared at each picture in disbelief, turning the pages slowly as he told me the story of what had happened on April 29, 1945. The unspeakable horrors caught on film, he said, were only a glimpse of what he had witnessed when he entered the camp.

In the years that followed, I came to realize that the events of that day had had a profound impact on his life, forever changing the way he would view the world and forming in him a steadfast resolve that good must always find a way to overcome evil.

Shortly after I graduated from college, he made copies of the photos for me and my siblings. Occasionally I would pull them out for friends who wanted to see a piece of history, describing the events as my father had described them to me. I found that the pictures, themselves a sobering remembrance of the Holocaust, became even more powerful when framed by the story of the men who had taken them. In having them published here for the first time, it is not my goal merely to present them for shock value, but rather to recount the sacrifice, courage, and honor of the American soldiers who liberated the camp, and to explain that these men—these liberators—were actually boys, barely out of their teens, who had survived the most historic battles of the twentieth century.

Even though I already knew the most important elements of my father's wartime journey, I interviewed him and his buddies extensively about their experiences before I began the process of writing this book, filling in gaps in the story as I tracked their progress across Europe. Wars, by their very nature, are extraordinarily complex events involving thousands and thousands of people. In order to clarify what could have become a confusing and dry narrative—and so as not to embarrass any of the good men of the 92nd Signal Battalion—I took the common artistic liberty of combining some characters and events and changing a few names when it was appropriate. This helped simplify the story without sacrificing the integrity, emotion, or adventure for the reader. The result was a moving (and often surprising) account of what it was like for the American soldiers in World War II—of how it felt to be fighting a war so far from home, uncertain of their fate and, more often than not, unsure of their true mission.

Though they were ultimately victorious, the war took an emotional and physical toll on the young soldiers. The photograph of my father on the cover of this book was taken in Salzburg, Austria, one week after the liberation of Dachau. He was twenty years old. Behind his pleasant features and slight smile, his eyes seem to hold the weight of what they had witnessed only days before, echoing a quiet sadness that, even to this very day, can still cause them to well up with tears.

Joe Sacco was the only child of immigrant parents. He worked on the family farm and, like many of the young soldiers of World War II, had never been away from home before being drafted. He had never held a weapon more powerful than a BB gun. He had never witnessed violence more intense than a schoolyard fight. And the first beach he ever saw was Omaha Beach at Normandy. His is the story of how an innocent farm boy from Alabama left home, trained with the U.S. Army, fought his way through Nazi-occupied France and Germany, and eventually helped put an end to the Holocaust.

He told me that for several years after the war, he would not speak about the atrocities he'd seen. He didn't think anyone would be able to fully understand the magnitude and significance of what he and his buddies had experienced. His eventual decision to show me and my siblings the photographs of Dachau, therefore, was not made lightly. He knew that the images were frightening, but he thought it would be important for his children to see what he'd witnessed first-hand so many years before—the shocking cruelty that had taken place in the Nazi concentration camps. And he wanted us, through his witness, to stand vigilant against such inhumanity ever being allowed to happen again.

Now, as his story is told, I share that responsibility with you.

Jack Sacco
Los Angeles

Sergeant Joe Sacco, Salzburg, Austria, 1945–one week after the liberation of Dachau.

Prologue

Dachau Concentration Camp

Dachau, Germany
April 29, 1945

The gun battle didn't last long. In about fifteen or twenty minutes, it was all over. Duthie and I had run up to the outer wall of the camp and were crouched down trying to figure out what the hell was going on. We had been told that this place was a prison, so we assumed there must have been some of the enemy holed up in there, trying to use it as a fortress. All I knew was that whoever was inside didn't put up much of a fight. Then again, by this point in the war, we were accustomed to kicking ass, and the Nazis were accustomed to hauling ass when they saw us coming.

Some guys were yelling that they were opening the perimeter gates, so I waved to Averitt, Spotted Bear, Chicago, and the others to follow us in that direction.

"Well, hell," Duthie said as we stood up. "Is that it? That was easy."

"Yeah, I guess so," I said as I picked up my gear and started to move out.

"How many krauts you think they took prisoner in there?" he asked.

I scanned the compound, trying to get a sense of the size of this place. From where we were, it looked like a well-fortified military post. A high brick wall and a moat surrounded the entire facility. Just beyond the wall, which was lined with barbed wire and punctuated by a series of guard towers, I could make out what looked like military barracks. I couldn't tell much more from the outside, other than the fact that it was big. "I don't know," I answered. "Big place. Bunch of 'em probably holed up in there like rats. The more the better. Assholes. I'd like to get my damn hands on a couple of 'em."

The other guys had come up alongside us as we rounded the corner near the main gate to the camp.

"What the hell's that smell?" Averitt asked.

"That's what Nazis smell like when they're scared shitless," Chicago answered.

Five U.S. infantrymen were standing guard at the entrance. We would have expected them to be happy about taking such a big objective with relative ease. They'd heard what Chicago had said, but they didn't laugh or even smile. They just looked down.

Duthie and all the men got real quiet as we crossed the threshold into the camp. The smell that had caught our attention as we approached the wall was now overwhelming. Within a few steps, we all came to a stop and looked around in disbelief.

I had seen carnage at Normandy. I had seen my buddies die right in front of me and whole towns get blown to hell. But I had never seen anything like this. Nothing could have prepared any of us for the horror laid out before our eyes. No one said a word. It was as if everything slipped into slow motion, where every second took an hour to pass, where every minute was filled with such incredible sorrow that it seemed it would never end. I was nauseated, dizzy, confused. My brain couldn't comprehend what my eyes were seeing.

Newly inducted Private Joe Sacco at Fort McPherson, Georgia, February 1943.

PART I

From Farm Boys to Soldiers

These are the heroes. They have come from the farmlands and cities across America to engage in the magnificent experience of battle. They come to defend their homes and loved ones. They come for their own self-respect, because they would not want to be anywhere else, even if it means giving up their own lives. They do not fear death. These men are real Americans, and as such, they are winners. By God, I love these men.

—GENERAL GEORGE S. PATTON

1

The Journey Begins

Birmingham, Alabama
October 1942

In all my memories of the farm, there is one day I remember the most. I suppose it's because I learned more about myself in the time it took me to read a simple letter that day than I had in the previous eighteen years of my life. When I looked up from that page, I realized that life is not a given, it's a gift, and that a man's destiny can lead him far from the home and the family he loves into places he never knew existed. Perhaps in all my years there, this was the first time I actually made an effort to remember the many people and things that had surrounded me for so long. Perhaps it was the first time I really took stock of what I had and realized its truest value. For whatever reason, it was the memory of that day—and the thousands exactly like it that I had allowed to go unnoticed—that would carry me through the years and the journey ahead. And for that I have been grateful every day since.

It was the afternoon of Friday, October 16, 1942, and the heat and humidity of a prolonged Southern summer was finally giving way to the welcome freshness of autumn. The nearby hills, covered with trees and just beginning to sparkle with the colors of fall, draped themselves down until they blended softly into the fields of the farm.

Along the western edge of our land—and occupying the flattest portion of the valley—was the small airport serving the town of Birmingham, Alabama. In earlier years, when we were kids, my cousins and I used to stand at the fence for hours watching the planes take off and land. They seemed magical to us, so big yet somehow able to fly, their noisy engines announcing their arrival, their wheels kicking up smoke as they touched the runway. Sometimes we'd even try to throw rocks at them as they came in for a landing, not out of malice but because kids on a farm try to throw rocks at most everything, and a low-flying plane was just too hard to resist.

Now, years later, as we went along with our duties in the fields, we barely seemed to notice the buzz of the machines flying overhead and landing nearby. For the most part, the soothing sounds of nature abounded here—the mooing of a cow, the occasional bark of a dog, the gentle breezes rustling through the trees as an old tractor hummed its way through the neatly groomed rows of corn, squash, tomatoes, cabbage, and other crops.

This rustic scene was, of course, musically enhanced by the ever-present sound of my uncles singing Italian songs. Each one apparently considered himself to be a great opera singer, each constantly trying to outdo the others in pitch and volume. Fortunately, they all seemed to be able to carry a tune, so on the rare occasions when they would harmonize instead of compete, it gave the farm the feel of a movie—sort of an Italian farming movie.

Papa and Mama had come from Sicily in the early 1920s. They sailed into New York Harbor before moving on to Chicago, where Papa began working at a factory. Before long he was offered a job by a childhood friend from Sicily named John Costa, aka John Scalice. Costa wanted Papa to drive a car for another Sicilian: a man named Al Capone. Papa refused. Capone was said to have felt disrespected, so he sent Costa to try once again to convince Papa to accept. Costa, by the way, was rumored to have been one of the hit men in the St. Valentine's Day Massacre. But Papa, being a decorated veteran of World War I in the Italian army and a man of considerable bravado, told Costa to tell Capone to shove it. It was just in keeping with Papa's personality to piss off the most powerful Mafia chieftain of the twentieth century. Which is exactly what he did. Which is exactly why we moved. Papa had heard talk of jobs and land in Alabama, where the weather closely resembled that of Sicily and the Mafia didn't exactly have a stronghold. And so to Dixie we came.

Upon our arrival Papa immediately learned one fact of life in the South—Italians were not welcome there. Alabama didn't have a Mafia, but they did have rednecks and the KKK. Neither liked anyone who wasn't a hick. So they designated certain areas of town—the run-down areas—as the places where blacks, Jews, Italians, Catholics, and anybody non-WASP had to live. Papa therefore bought a farm on the eastern edge of Birmingham, adjacent to the airport and near the farms of fellow Sicilians John Musso, Joe DiGratta, Tony Sciatta, and Mike Renda.

Now, along with an extended family that eventually included grandparents and a large assortment of aunts, uncles, and cousins, we worked the land until it yielded a rich harvest. We had settled in the South, but our language and customs were Italian. Thus, we held paramount in our hearts the two traditional sources from which we drew our daily strength: faith in God and love of family. These values, along with a willingness to work long and hard, saw us through the difficult times of the Great Depression and brought us even closer together.

Now, just four days after my eighteenth birthday, this Friday after-noon seemed entirely typical. We had worked in the fields all day and had just finished loading the truck with vegetables for the next morn-ing's delivery to the farmers' market downtown. The sun was complet-ing its long journey toward the distant clouds in the west, and a cool, refreshing breeze silently swept through the valley. The smells of sup-per cooking began to rise from the house, calling us home for the quiet of evening. This was, as far as I was concerned, that most per-fect time of day.

All of us—aunts, uncles, cousins, and grandparents—lived in what might be described as one big house with different sections. And though I was technically an only child, I was never really lonely, because there were always plenty of other kids around. In the evenings we would gather to eat, tell stories, sing old Italian songs, and, of course, engage in animated arguments. When the weather was nice, like it was that day, we would eat outside. These evenings were the good times, the best times, because we were happy, and we were together.

As the men washed up for dinner and the women prepared the table beneath the giant oak tree beside the house, Mama approached me with a letter. "This came for you today," she said. "It's from the government."

I dried my hands, took the letter, carefully examined my name on the front, and then opened it.

"Your friends and neighbors have selected you to serve in the Armed Forces of the United States of America to defend our country." It was signed Franklin Delano Roosevelt, President of the United States of America.

Mama took the letter from my hand and read it aloud to the gathering family. "Joe's going into the Army," she announced slowly as she began, translating into Italian as she went so that my grandparents could understand.

I remember walking the few feet from the oak tree to the fence that bordered the fields. I could hear her reading, but my eyes and heart became fascinated with the scene laid out before me. Colorful ribbons of scarlet and orange were beginning to stretch themselves across the deepening sky, causing the crops below to shimmer as though they were painted with sparkles of silver, gold, and red. These glistening fields and the rolling hills beyond looked exactly as they had on thousands of other evenings. But somehow they looked different. Somehow they looked more beautiful than I had ever remembered.

Grandpa Sacco approached and put his arm around me. *"Dio ti proteggerà, Giuseppe. Dio ti proteggerà."* God will protect you.

"Si, conosco," I said to reassure him. I know He will.

I looked back at my family, who were by now in the first stages of being seated at the table. Papa and my uncles were discussing the politics of the war as Mama, Aunt Mae, and Grandma Amari were bringing some platters of food outside from the kitchen. The teenage cousins, having worked with us in the fields all day, were somewhat quiet, while the younger kids were bustling about, as they did every evening when it was time to eat. As I watched, I came to realize that everything I knew, everything I cared about, was in this one scene. I took in a deep breath, smelling the warm aroma of supper carried on the fresh country air, drinking in the scene that surrounded me, hoping that I would never forget this smell, these sounds, this love, this moment.

That same letter was delivered to hundreds of thousands of other boys throughout the nation. Each of us began a journey that day—a journey that would lead us into the greatest conflict in recorded history. For my part I had read the newspaper accounts about the brewing war in Europe, but I really didn't have much of an idea of what awaited me. All I knew was that I loved the United States of America, that my country needed me, and that I would serve her, if need be, with my life.

Birmingham, Alabama
February 24, 1943

The bus station was only about three miles from the farm, but on the morning I left home, it seemed like it was on the other end of the earth. All the relatives and friends from the surrounding farms had come over the night before to say good-bye and to wish me well. I had a few cousins who would probably be drafted within a year or so, but I was the oldest and I was going first, so I guess it was a big deal for everybody to come by and see me before I left. I enjoyed it and naturally didn't want it to end.

Now, after an abbreviated night's sleep, I was alone with Papa in the produce truck riding to the station. It was quiet in the truck. There were thousands of things I wanted to say, but I felt too many emotions swirling around to make much sense out of any of them. The one thing I knew I felt was fear.

As we pulled into the station, Papa spoke. "Joe, you afraid?"

I wanted to tell him, *"Hell yes!"* I wanted to tell him that I didn't know if I'd ever see him or Mama again and that the thought of it scared me to death. I wanted to tell him to turn the truck around and take me home. He looked at me with those piercing blue eyes of his and gave a little smile.

"No," I heard myself saying.

"Okay, good," he said. "Good."

Just before I boarded the bus, he traced a small cross on my forehead and then embraced me. *"Padre, Figlio, Spirito Santo. Dio ti benedica e ti accompagni sempre."* May God bless you and be with you always.

"Everybody get on the bus!" the driver yelled out. "Time to roll."

I hopped on board and found a seat. All the boys were quiet as the bus pulled away. All of us were looking back into the parking lot where our fathers stood, as if in these last few seconds of visual contact we might gain some strength, some courage or knowledge to help us on the road ahead. I sat next to a guy I didn't really know but whom I'd seen around town. He was an Italian whose father owned a grocery store in Woodlawn, which was a predominantly black neighborhood near our farm. I recognized his father standing next to Papa as we pulled out.

"Hey. Joe Sacco," I said, extending my hand.

"Yeah," he said, "Tony Palumbo. I've seen you around before."

"Yeah, me, too," I said.

"Your papa is Mr. Jake Sacco," he continued.

"Yeah." I was surprised. "Say, how did you know that?"

"Because Mr. Jake delivers produce to the store on Tuesdays."

"Oh, yeah? I didn't know that," I said.

"Oh, yeah," he said, smiling. "Your papa and my papa have known each other a long time."

In a way I felt better, a little more reassured, a little less alone than I had a few minutes earlier. Conversations on the bus were picking up. Some of the boys were loud and made a point of laughing and telling stories, I guess by way of showing that they weren't afraid. Others just sat there quietly and looked out the window.

I turned to Tony. "So, you going into the Army or what?"

"Oh," he began, "well, after basic training I think I'll go into the Army Air Corps."

"Really?"

"Oh, yes. You see, I like the idea of being able to fly over all the fighting and drop bombs and then fly away," Tony laughed.

"Not me," I said.

"What are you gonna do?"

When I had taken my test back at the draft board in October, they'd given me the choice of joining the Army, the Army Air Corps, the Navy, or the Marines, so I had already thought this out pretty carefully.

"Okay," I said, "here's the deal. I figure, can I run from a sinking ship? Not unless I can figure out a way to walk on water. I mean, I'm a good guy and all, but I ain't exactly Jesus. Okay, flying over the trouble is nice, but, hey, the damn Germans have these big old guns that can knock a big hole in the side of a plane. So can I run from a plane that just got shot down? Nope. Nobody can. And I can't fly."

"But you can have a parachute," Tony corrected.

"Yeah," I continued, "that's great. A parachute. So you can just float to the ground, assuming you live through the part about the plane exploding, and all the way down you're this big-ass target for every Nazi in Europe. And then let's say you do make it to the ground without getting shot a dozen times, then you still don't know where the hell you are, because your ass has been flying in a plane for the past two hours, so you're turned around all kinda ways, and it's probably night, and you start running, but you don't really know which way to run, and you might run smack into Hitler taking a piss, and he shoots you and then finishes his piss."

Tony looked kinda stunned.

I continued. "Okay, so that meant no Navy and no Air Corps. Okay, when I took my test, they had this Marine there, and he was all dressed up in this Marine dress uniform. Boy, it looked nice. He had a sword and everything, all polished up. But, hey, I figured, we're gonna be fighting a war, so when are you gonna wear that uniform? When you're dead. So I chose the good old Army. You stay on the ground, you can run, hide up under stuff, get away—whatever you have to do. It's not that I'm scared or anything like that, because they're gonna be training us to fight, but I figure that if I gotta run to stay alive, then I want to be able to haul some ass."

Tony looked out the window, thinking.

All the boys got quiet as the bus pulled into Fort McClellan, Alabama, and rolled to a stop. The driver opened the door, and we all sat expectantly as a very muscular, energetic soldier sprang aboard.

"Get off the bus! Let's go! Time to move! You're wasting my time! Get off the bus right now!"

We scrambled as fast as we could. When the last boy had exited, the driver shut the door, and the bus slowly pulled away. We must have looked like a ragtag group of guys, all wide-eyed, not knowing what to expect, not sure what to do or even how to stand. The drill sergeant, neatly groomed, with perfect posture and a serious attitude, surveyed the scene and then spoke. "Okay, my name is Sergeant Turner. You address me as Sergeant Turner. When I ask you a question, you say, 'Yes, Sergeant Turner' or 'No, Sergeant Turner.' Is that understood?"

There was quiet among the ranks.

Sergeant Turner raised his eyebrows. "IS THAT UNDERSTOOD?"

"Yes, sir," we replied, in something as close to unison as we could manage.

"Now, what the hell did I just tell you? You do not call me 'sir.' You call an officer 'sir.' You call me 'Sergeant.' Now, is that understood?"

"Yes, Sergeant."

"Okay, that's better," Turner continued. "Now, gentlemen, welcome to the United States Army. This week you will be processed, given your medical exam, and sworn in. You will be briefed on Army procedures, protocol, et cetera, et cetera, et cetera. I will be with you if you have any questions. Do you have any questions?"

We looked at the sergeant without uttering a word.

"No questions," he said. "Outstanding! It's going to be a beautiful day in the U.S. Army!" He looked at his watch and then said, "Okay, in seven minutes we'll get started with your medical examination. Until that time I want you to start policing the area. You know what 'policing' means? It means to go around and pick up any trash you find, that's what. So start policing this area! Let's go!"

All of us remained motionless. Here's why: The driver of the bus had told us that the drill sergeant would try to make us pick up the trash, but that nobody could order us around until *after* we were sworn in to the Army. He had said, "When he tells you to pick up the trash, just stand there. If you do what he says before they swear you in, you'll get in trouble, and he'll think I dropped off a load of bumpkins."

So now Sergeant Turner was yelling at us to police the area, and we were just standing there looking at him like a load of bumpkins. I was thinking, *Oh, crap! I hope that fella knew what the hell he was talking about, because he's gone now, and I'm standing here looking at this unfriendly sergeant, and I don't want to get my career off on the wrong foot.*

He glared at us for a few minutes, but once he realized we were onto him and that we weren't going to pick up any trash, Sergeant Turner led us into the big white wooden building behind him. We were then directed into a large room set up for medical examinations. This room was like a maze, with doctors and their equipment situated so that we would have to go from one to the other to the other for different evaluations before we would be allowed to exit the room on the other side. Sergeant Turner had one last thing to say to us before he left: "Strip and wait until the doctor calls your name."

These boys might have claimed to be brave so long as we were on that bus, but I can guarantee that none of them wanted to take off their clothes. At least I hope none of them did. I know I didn't. We all stood there, looking around as if we'd forgotten what he just said. The sergeant was impatient. "Let's go! Strip right now and get in line!"

All the boys slowly took off their clothes, starting with shirts and shoes and things that didn't matter that much. When we got down to the part about taking off our pants and shorts, we just did that real fast and then quickly stood upright again. As far as I could tell, everybody was looking straight ahead and not talking. Finally this one guy started laughing, and then we all laughed. That eased up things a bit, but I still didn't want to make eye contact with anybody.

I didn't know whether I was more embarrassed due to the fact that I was standing there naked or because I was surrounded by forty other guys standing there naked. Either way it wasn't exactly a pretty picture.

We had to go from one doctor to another, and they checked everything from the top of our heads to the bottom of our feet. And I mean everything. The experience was *not* fun. In fact, I felt like slapping the hell out of a few of them. Back home, people would get mad and *say* they were going to shove something or the other up your ass, but they don't actually *do* it.

After what seemed pretty much like forever, we were allowed to redress and were sent into an adjoining hallway that was lined with benches. A soldier told us to wait there until our examinations were processed. I think all of us felt good to have our clothes back on.

"Hey, which one of you is Sacco?" an officer with a clipboard yelled as he walked quickly through the area.

"Right here," I said, raising my hand.

"Okay," he said, looking carefully over the chart in his hand. "Just one thing. We don't have a birth certificate for you. You got one?"

"Uh, no," I said, digging into the empty pocket of my shirt, as if some phantom birth certificate might suddenly materialize there.

"Well, we need one," he said.

"I . . . uh . . . I don't think I have one. I've never seen one . . . never heard anybody say anything about one. Maybe the doctor back home has it."

"We need one for proof of birth," the soldier said.

"Hey, I'm standing right here," I replied. "Ain't that proof that I was born?"

The other boys started laughing. The officer said, "Okay, pipe down. We'll get it straight later," and left the room.

I guess they eventually believed that I had been born, because they accepted me into the Army. With regard to my medical exam, I was rated 1-A. One of the boys on my bus was turned down for medical reasons. He seemed disappointed, but I told him, "You're lucky. I wish I could go home." A lot of the other boys felt the same way.

After a few minutes, Sergeant Turner came through and led us into a small room where we were formally sworn in to the United States Army.

Fort McPherson, Georgia
February 26, 1943

Probed, poked, measured, grabbed, tested, and newly sworn in, we were put aboard a bus for Fort McPherson, Georgia, where we were to get our shots and take written aptitude exams. I don't know why we had to go all the way to Georgia for those activities, but it's not like we had a choice, so there we were.

We had to stand in a long line—the Army is fond of long lines—in order to get our shots. The boy right in front of me kept prattling on about how he was afraid of needles.

"Oh, come on," I told him. "Quit being such a squirrel."

"I'm n-n-not a sq-sq-sq-kidding," he said "Those d-d-d-damn things h-hurt like h-h-hell. I d-d-don't l-like 'em."

At this point I was wondering if they had a shot to make him stop stuttering, because I found that the more he talked, the more I wanted to slap the words out of his mouth. Anyway, I figured that since they probably wouldn't let me slap him, I would try to make him less nervous.

"Look up there," I said. "See all those other boys?" I didn't have the patience to wait for him to respond. "All of them are getting their shots. And look, it's no big deal. Nothing to be afraid of."

"Well, I d-d-d-d-don't know."

We eventually got up to the front. He took his shots, turned around, and walked off.

"Hey, th-th-thanks, J-J-Joe. It's not so b-b-b-bad."

I was up next. I took a shot in each arm, passed out, and hit the floor like a rock. When I woke up, I realized they had dragged me over to one side. Some of the boys were still in line taking their shots and then stepping over me as they left.

Stutterbutt was there. "Hey, J-J-Joe? You okay?"

I jumped up as fast as I could. "I'm okay," I said loudly. "Everything's okay."

The boys in line started laughing.

A nurse came over, gave me a glass of orange juice, and made me sit in a chair until the dizziness went away. I talked to her until all the other boys had completed their shots. It wasn't that she was all that pretty or anything, or that I was all that dizzy, it's just that the longer I sat there, the more orange juice she gave me.

Later that afternoon we went over to the supply building, where we were issued shirts, pants, socks, shoes, caps—everything. The supply sergeant seemed slightly psychotic. One of the boys in front of me must have said something wrong to him, because he jumped up on the counter and began swinging a big knife. He laughed maniacally and yelled, "You know what this is?" The boy said, "Yeah, it's a knife." And the supply sergeant screamed, "No, you dumb-ass, it's a machete!"

When I got to the front, I handed him my papers. He stacked my uniform on the counter without a word, and I went on my way. As I was leaving, I heard him say, "Joe Sacco . . . wop."

I stopped in my tracks. One of the other boys near me put his hand over his eyes, looked down, and whispered, "Oh hell."

I heard Sergeant Turner, who was standing near the door. "Sacco." He shook his head and cracked a slight smile. I looked at him for a moment, and then I thought, *This crazy-ass supply sergeant's not worth me getting in trouble to start off my career. And besides that he's got a big knife.*

As I walked past Sergeant Turner, he said, "Good discipline, soldier. Now let's go get dressed."

And get dressed we did—but first we had to get our hair cut. Sergeant led us into a room set up like a barbershop, except there were mirrors on only one wall, the one behind the chair, and there were no magazines. There were maybe eight barbers cutting hair as fast as they could. Each haircut took about fifteen seconds.

The guy in front of me in line kept combing his hair while we waited. When he sat down, the barber said, "So how would you like that done?"

The boy said, "Well, I suppose take a little off the side and in the back and just trim it up."

The barber said, "Okay." And zip, cut off all the boy's hair. He spun him around to face the mirror, and the boy screamed. I thought that was funny as hell.

When it was my turn, I had to go to the same barber. He pulled the same routine with me. "How you want it?"

I said, "Well, hell, just cut it off."

He said, "You don't want anything special? 'Cause I can do anything you want."

"You can?"

"Oh, yeah," he said. "I have the authority."

I said, "Well, what if you just make it look real neat? Uh, *not too short.*"

He said, "Oh, okay, I can do that."

Fifteen seconds later he spun the chair around to face the mirror. I was looking at a smiling barber and a bald-headed me. There was this initial split second when I wondered if the reflection was really of me. I reached to touch my head. The barber yanked the apron from around my neck and pushed me out of the chair. "Play with yourself on your own time, kid," he said. "Next!"

After that, Sergeant Turner showed us the proper way to put on our uniforms. Now we felt like soldiers. At least we looked like soldiers.

The next day we were taken into a large room and given a series of written tests. I must have scored well on telegraph communications, because they told me that after basic training I would be assigned to the 92nd Signal Battalion. I didn't understand exactly what that meant, but I was quickly learning that I wasn't required to understand everything. In fact, Sergeant Turner had told us that the only thing we really needed to understand is that we had to follow orders. And my orders were to board a train later that night bound for Camp Crowder, Missouri, where I would report for basic training.

I packed my duffel bag, which the crazy supply sergeant had given me along with my uniform, and boarded the troop train. We had to pass right through Alabama on the way to Missouri, but it was already deep into the night, and by the time the train came to Birmingham, I had fallen asleep. Not that I could really have seen anything from a moving train in the middle of the night, but I still felt like I had missed my last look at home for a long time.

2

Basic Training

We traveled all through the night and well into the next day before we pulled into the Ozark Mountain town of Joplin, Missouri. It was cold and rainy outside. I could see four sergeants waiting on the platform, barking out orders as the train rolled to a stop.

"Get off the train! Let's go, party's over! Move it!"

Like every man on board, I grabbed my duffel bag and ran toward an exit.

"When you hear your name called, you will respond and you will get into the appropriate vehicle. You're in the Army now, so let's move!" the sergeants ordered.

I heard my name being called at the other end of the platform. "Sacco! Thompson! O'Brien! Haul some ass!" I picked up my bag and, as instructed, hauled some ass in that direction. I reached the sergeant and said, "Sacco." He marked his clipboard and said, "Hop in, soldier."

I tossed my bag onto the back of the troop truck and jumped in after it. Once the vehicle was loaded with about ten boys, the sergeant slammed his hand on the driver's door and said, "Good to go! Outstanding!" Before we could pick up speed, he had run around, opened the passenger door, and hopped in.

The truck pulled out of the station and rattled along a bumpy road that led up into the Ozarks. None of the boys said a word as we rode, shivering in the cold, damp air. After fifteen hours of rocking inside a relatively warm train, this was a brisk wake-up call. We seemed to be going higher and higher into the hills, and it was definitely getting colder as we went along. After about twenty minutes, we drove through the gate of Camp Crowder, pulled into an open area, and creaked to a stop. The sergeant hopped out from the passenger's seat and came around to the back of the truck.

"Avery and Sacco! End of the line! Let's go," he said, looking at his clipboard. "Move it, boys. We ain't got all day!"

I grabbed my belongings and jumped out. Another boy followed me.

"All right. Sacco. Avery." The sergeant checked our names off the list. "I'm Sergeant Gray."

"Averitt," the other boy said.

Sergeant Gray looked at the list and then at the boy. "You say something, soldier?"

"Uh . . . name's Averitt. Not Avery," he said.

The sergeant checked his list again. "Congratulations, Avery. Okay, men, this is where you live. Welcome home," he said, pointing to the building behind him. Then he turned to the driver. "The rest go to H-6 through -8. I'll be there in five. Get 'em off the truck and wait for me." With that, he slammed his hand on the side of the vehicle again and yelled, "Good to go!"

The truck jerked and sputtered and slowly pulled away, leaving the three of us in front of a rather stark-looking white two-story wooden building. Next to it stood a duplicate of itself. Next to that one, another, and then another. Fifteen or twenty in all. The only thing that distinguished one from the other was a number painted in black over the door. The sergeant had said, "Welcome home," but this place didn't look anything like my idea of home.

"Hey, I'm Paul Averitt," the boy next to me said, extending his hand. "Nashville, Tennessee." He was thin, sort of a wiry-looking little fella with light hair, green eyes, and a big smile.

"Yeah, Joe Sacco. Birmingham."

"No kiddin'?" he asked, grabbing my hand and shaking it vigorously. "Another Southern boy! Well, I'll be!"

The sergeant interrupted. "Hey, Avery!"

"Averitt."

"Yeah, whatever. You're in barracks H-3. This is H-4. You wait right here. I'll be back. Sacco, come with me."

The sergeant darted off toward the barracks. I quickly turned to Averitt. "Hey, buddy, nice to meet you."

"Let's get a move on, Sacco," Sergeant Gray said, already six steps ahead of me.

"All right, buddy," Averitt yelled out as I ran to catch up with the sergeant.

"You're upstairs," Sergeant Gray said as he bounded up the outside steps leading to the second story. As we reached the top, two enlisted men were coming outside onto the small porch.

"Well, what have we here?" one of them asked.

"This here is a brand-new soldier, boys," Sergeant Gray answered.

The other one, the taller of the two, smiled. "Hey, Sarge, this the one from Alabama?" he asked, intoning a fake Southern accent when he hit the word "Alabama."

"This is him," said the sergeant as he pointed the boys back inside the barracks.

"No shit? Alabama! Come on in, Bama," the taller one said, holding the door open.

I followed Sergeant Gray into the barracks. It was one large room with about forty single bunks spaced a few feet apart. The head and foot of each bed alternated, so that if one boy slept facing north, then the next one slept facing south. I suppose they had a reason for arranging it this way, maybe so that we wouldn't wake up looking directly into another man's face. That could be scary. But I also didn't know if I liked the idea of having some guy's feet near my face, not to mention the fact that I would have two asses, one on each side, aimed in the general direction of my head all night.

In any event, behind each bed, along the wall, were little areas where we could hang our uniforms. Along the center aisle, in front of the beds, were footlockers. At the very end were two other rooms; the left was the latrine, the right were the showers. About thirty or so guys were sitting around on the bunks, some talking, some reading, some writing letters.

"Listen up, fellas," the sergeant announced as he entered the room.

"Hey, guys, Bama's here!" the shorter guy from the porch said. All the men in the room turned and looked at us.

Sergeant Gray spoke. "Okay, this is Joe Sacco. He's from Birming-ham, Alabama." There were a few yelps and hollers when he said that.

"Calm down," he continued. "Show him the ropes, make him feel welcome, get him to chow, and get him situated." The sergeant turned and hurriedly exited the room.

"No prob, Sarge," said the tall one from the porch, who had a tattoo on his left arm of the name Lucy inside a heart. He walked toward me with a slight limp. "Hey, Bama, Sonny Macaluso. Chicago. Busted the hell outta my leg a few days ago. I'll survive."

"Joe Sacco." I pointed to the tattoo. "My grandmother's name is Lucy. Well, Lucia. It's the Italian version."

"No shit?" Macaluso said.

"Hey, look at this," one of the boys said. "Macaluso's girlfriend is Bama's grandma."

"So these are the guys," Macaluso continued over their laughter. "We heard you were coming. Most of us got here last week. That's Jones. He's from Toledo. Scully, from New York City—"

"Long Island," a voice, apparently belonging to Scully, interrupted.

"It's the same thing. Shut up," Macaluso deadpanned.

Scully rolled his eyes. "It's not the same thing."

A pillow came flying at Scully from a few bunks away. The boy who threw it said, "Shut up, Scully." Another one said, "It's the same damn thing." Scully caught the pillow and said to the boy who had thrown it, "Okay, I'm keeping this, and this time you're not getting it back."

"Don't make me come over there, Scully," the pillow thrower said.

Macaluso laughed and continued. "Greer from Providence, Rhode Island. Hey, Bama, you ever been to Rhode Island?"

"Uh, no."

"Don't worry. Nobody has. Just Greer and some other fella live there." Macaluso pointed to another boy. "That's Cook, from Jersey. Armstrong, Cincinnati or someplace. Right, Armstrong?"

"Cleveland."

"Yeah, Cleveland, Ohio." He went around the room, pointing out most of the guys and calling out their home states. A few of the boys just ignored him, but most played along. When he was all done, he asked, "Hey, any of you guys from the South like Bama here?"

One guy said, "I'm from southern Ohio."

Another one said, "That's not the South. I'm from Kentucky."

"There you go, Bama," said Macaluso. "Another Southern boy! Whataya think of that?"

I said, "All I know is that from Alabama I'd have to go north to get to Kentucky, so as far as I'm concerned, he's a Yankee."

The other boys laughed and said things like, "You're all right, Bama." Most of them came up and introduced themselves. They seemed to have lots of questions about Alabama. One of the guys asked if people in Alabama wore shoes. "Hell no," I said. "They don't have shoes there." I pointed at my boots. "To tell you the truth, these are the first pair I've ever had." They thought that was funny.

One of them, I think it was that Scully fella, said, "No shit?" and got a real serious look on his face. That made everybody laugh even louder. They all seemed to be good guys. After we talked for a while, they showed me the ropes about how to clean my area, how to pack my clothes away in military fashion, and what was expected of me. Macaluso helped me put my gear—things like a shovel, a pup tent, rainwear, a flashlight, water, C rations, and other necessary items— into my backpack.

We went to chow later that evening. They had served us good food at Fort McClellan and Fort McPherson, but the food here at Camp Crowder was even better: meat, potatoes, gravy, fresh vegetables, hot bread, butter, fruit, and plenty of milk. I suppose that if they expected you to make it through basic training in one piece, they had better feed you some decent food and plenty of it. I didn't know if I liked being in the Army, but I did enjoy chow time. I must have been especially hungry after all the excitement of arriving at the camp, because I ate like there was no tomorrow. Two helpings of meat loaf. More mashed potatoes and gravy. Another scoop of vegetables. Two more rolls smothered with butter. And more milk. Just keep the milk coming. In fact, I drank so much milk the other boys started to wonder if something was wrong with me. The guy from Kentucky said, "Hey, Bama, you're gonna make yourself sick. Save some for tomorrow!"

Macaluso was laughing, "What's the problem? They don't have food in Alabama? Take it easy, Bama."

Napoleon is quoted as saying that an army marches on its stomach, which conjures up an interestingly comical mental image, but the more I ate, the more I realized what he meant and the more I was willing to give Army life a chance.

After chow we had a little free time, and then, at nine o'clock, it was lights-out. I expected that I might have some trouble sleeping that first night, but I was so tired—and full—by the end of the day that I drifted off within seconds.

Early the next morning, at an hour the other boys referred to as "oh-dark-thirty," I was startled from my sleep by the sound of a bugle. It must have been over a loudspeaker outside, but it sounded like it was being blown right in my ear. It scared the hell out of me. Within seconds the front door of the barracks swung open, and a sergeant entered the room blowing a whistle.

"Time to rise and shine! Up! Move it! The day's slipping away, gentlemen! Let's move!"

I jumped from my bunk, a little confused about what to do first. One thing I knew for sure was that the floor was cold. My bed, on the other hand, was warm. I heard Macaluso's voice. "Hey, Bama, shower's this way. Hurry up! Should've taken one last night. Water might be kinda cold. Class C uniform. Be downstairs in fifteen." With that, Macaluso was gone, disappearing quickly into the showers.

Armstrong, from Cleveland, threw a sock at me and started laughing. "Macaluso's full of shit. Gimpy-assed son of a bitch uses all the hot water—*then* it's freezing. You better hurry, Bama."

I grabbed some soap and sprinted for the shower.

Washed and dressed, we met downstairs and walked together to the mess hall, where we had a full, hot breakfast. Sergeant Gray, who was at our table and had apparently heard about my performance at dinner, said, "Hey, Sacco, try not to eat everything in the building. They might make you exercise a little today, and I don't want you upchucking all over this U.S. Army training facility." The other boys laughed, but I didn't care. I had eggs, pancakes, bacon, sausage, toast, and two big glasses of milk.

After breakfast we lined up outside the barracks to begin the day's training. The sun was just beginning to rise, but it was cloudy, and I could tell it was going to be a cold, damp, dreary day. A small sergeant, trailed by a tall corporal with a clipboard, addressed us. This was the one who had run through the barracks earlier this morning blowing a whistle.

"Okay, there are some new men among you, so in case you don't know me, I am Sergeant Youngblood. This is Corporal Dodge. I am here to train you to be soldiers in the United States Army. I don't care what you think you can or cannot do, because you are going to do what I say, when I say it, and how I say it. This is my army, and you are my men. For the next ten weeks, I am responsible for you and for your training. If you're properly trained, then that means you will survive to tell your grandchildren about it. If you're not properly trained, you will get your ass shot off by some son of a bitch who *is* properly trained. Are there any questions?"

We looked at Sergeant Youngblood without uttering a sound.

"Good," he continued. "Now, I'm looking at you and wondering, 'What's it gonna take for me to get you ladies ready for a war?' Corporal Dodge, what the hell is the U.S. Army sending me?"

Dodge looked up from his clipboard. "I don't know, Sergeant."

"I don't know if I can work with girls like these!" Youngblood continued. "I need the Army to send me men, real men, so that we can go fight a war! And stand at attention!" He jumped toward one guy and yelled, "You even know what attention is?"

The boy stood motionless until the sergeant moved away. "Look at these pitiful sons of bitches they send me, Corporal Dodge. I need men to fight a war. What am I supposed to train these fellas to do, pick turnips? We ain't got no turnips to pick. We got some Nazi ass to kick. There's a difference."

One of the fresh arrivals must have smiled, because Youngblood was in his face. "You hear something funny? Because I don't remember saying anything funny! Are you laughing at me, soldier?" He was yelling at top volume, and the veins were popping out of his neck.

The smile quickly vanished from the boy's face. The sight of it almost made me laugh, but I sure didn't want Youngblood screaming at me. The sergeant continued, "Where you from, soldier? You fresh from the farm?"

The neophyte soldier tried to answer in military fashion. "No, sir!"

"Who are you calling 'sir,' soldier? You see an officer around here?"

The soldier stammered, "Uh . . . you."

"Corporal Dodge, why the hell was this soldier not briefed on proper U.S. Army protocol?"

"I don't know, Sergeant Youngblood," the corporal answered without emotion, as if he had heard the question dozens of times.

Youngblood turned his attention back to us. "I know all of you boys miss your mamas, but I'm gonna need you to get your heads out of your asses long enough for me to turn you into soldiers worthy of the United States Army. Is that clear?"

There was silence.

"IS THAT CLEAR?"

"Yes, Sergeant!" resounded from the assemblage.

Youngblood clearly loved this. "Okay," he said. "Now go get your gear. We're going for a walk."

I ran upstairs with the other guys and grabbed my backpack. What I hadn't realized when Macaluso and I had packed it the night before was that it weighed about fifty pounds. Fortunately, I was accustomed to carrying fifty-pound sacks of grain around on the farm, so I tossed the pack on my back, jogged out the door, and got into formation.

Sergeant Youngblood had us face right. He then came up beside us and said, "Move out." We formed two columns, one on each side of the road, and started walking. It was pretty quiet at first, the only sounds being those of the men walking and the gear rattling around in their backpacks. Then, after a few minutes, I could hear guys starting to breathe a little harder and could see the steam coming from their mouths in the cold morning air. I noticed that whenever I took a deep breath, the coldness would reach halfway down my throat and sort of pound the inside of my chest. That would have been the most unpleasant aspect of the hike if it hadn't been for the backpack, which started to dominate my attention after about a mile and which seemed to be getting heavier and heavier as we went along.

Back on the farm, I would carry a heavy sack of feed, but I'd carry it only as far as the barn, and then I'd put it down. If it got uncomfortable, I'd simply shift it to the other arm. But there aren't a lot of ways to shift a backpack around if one still expects to walk. And Sergeant Youngblood apparently expected us to walk. And walk. And walk. After a couple of miles, he started singing, and the men answered. He'd say, "You had a good home, but you left."

And we'd say, "You're right."

"You had a good home, but you left."

"You're right."

The singing helped us know which foot we were supposed to be on, left or right, and supposedly helped us forget the distance, but I was still getting real tired, and I honestly didn't know if I could make it. We ended up walking twenty-five miles. I had always considered myself a pretty strong guy, but the farther we walked, the more I kept having this image of myself falling out dead along the way. Some of the guys couldn't make it, and a jeep had to come pick them up. It was getting dark by the time we approached the barracks, and I had to use every ounce of energy I had in order to take the final steps. Fortunately, the thought of supper was very appealing, so once we got within smelling range of the mess hall, I felt a resurgence of dedication and strength.

We started doing this almost every day, getting up before sunrise, hiking all day, and getting back just before dark. After a while it wasn't so bad. Usually the sergeant led that "left and right" song, but sometimes we'd sing things like "Three Blind Mice," "The Battle Hymn of the Republic," or whatever came into our heads. On the days we didn't hike, we would have an early-morning two-mile run before our physical training, at which time we would generally climb ropes, scale walls, and negotiate obstacle courses. As might be expected, we were taught how to properly handle and fire a weapon. We were also trained in hand-to-hand combat and on the finer points of tossing hand grenades and other ordnance. All in all, we put in a full day's work. And this was before lunch.

They kept us busy doing some things that seemed to make no sense, like repainting the barracks, putting up a fence and then taking it down, or carrying a big rock across a creek. There's an old Army saying that goes, "It's not for us to reason why, it's just for us to do or die." Sergeant Youngblood told us that some guy had written that in a poem a long time ago, but that he had actually gotten it from the armies of old. For the sake of my own morale, I found it was better not to wonder about why we did anything or where the saying originated, but just to do what I was told and then go eat.

The training I enjoyed the most involved shooting at the rifle range. I was, if I may say so myself, pretty damn good. When we were kids back on the farm, we would take a BB gun out into the woods and shoot cans and squirrels. So as long as we were shooting at targets on the shooting range, I was having fun. I didn't know how I would feel about shooting another man, even if he was considered to be an enemy. Then again, my attitude was that if it's between him shooting me or me shooting him, then he needs to be ready to meet his Maker, because I'm trained to put a bullet between his eyes now and work it out with a priest later.

On the day we took our final shooting test, I was lying on the ground firing away, hitting the bull's-eye every time. This little fella named Duncan who was shooting next to me said, "Hey, Sacco, how the hell are you doing that?"

"How the hell am I doing what?"

"How are you hitting that target every time?"

"It's because I'm good as hell, that's how."

"I can't hit the damn thing," he said.

I looked over and saw that he had hit the target only once or twice and never came anywhere close to the bull's-eye. "What the hell are you aiming at?" I asked.

"Must be something wrong with this damn rifle," he said.

"Yeah. Maybe there's something wrong with you."

"Come on," he whined. "Help me out, buddy!"

I said, "All right. Watch this. And don't shoot." I'm left-handed, and he was on my left side, so I started shooting one round at my target, then one round at his. I'd hit mine, I'd hit his, then mine, then his, over and over again. And I was hitting the bull's-eye on both most of the time. The only problem was that it involved me moving my aim back and forth and back and forth, which disrupted my concentration and caused me to miss on a few shots altogether. When it was all said and done, we both ended up getting rated "Marksman." Of course, he hadn't shot squat. I knew that if I had just shot at mine, I would have been rated "Expert." I didn't mind. In fact, if I'd done too well, the Army may have wanted me to be a sniper or sharpshooter, which didn't sound like a lot of fun. Besides that, I figured the less often I had to go tell a priest that I killed somebody, the better off I'd be.

There were two main activities that marked my day, the two I looked forward to the most: meals and mail call. We were served three great meals per day . . . well, two plus C rations if we were hiking. C rations were cans of cold food manufactured specifically for the Army and tasting specifically like cans of crap. I couldn't discern if it was physiological or psychological or some combination of the two, but after hiking for seven hours without authentic edibles, a can of C rations started to become strangely and remarkably appetizing. At first I didn't think I would ever get used to the taste. In fact, I didn't *want* to get used to the taste. Then, oddly enough, not only did I grow accustomed to it, I found that I actually started looking forward to opening a can and consuming the food substitute that lurked within.

Fortunately, we were almost always back in the mess hall for supper, and that's where I would feast. The menu was different each night, but the food was always plentiful and good. I know each of us had worked up a big appetite through the day, but this meal was more than just a refueling stop for me. This was the part of the day—the only part of the day—that even remotely reminded me of home. And as long as I ate and talked to the guys and laughed and enjoyed this meal, I was able to temporarily forget where I really was and what I was really doing.

The meals were good, but everybody's favorite time of day was mail call. Every afternoon a jeep would pull up and we would gather around one of the soldiers assigned to deliver our mail. Mail was like gold. If we got it, we were happy. If we didn't, we were jealous of those who did. It was that simple.

My parents and all of my relatives must have understood the importance of mail call, because they sent me lots of letters. Mama would write and give me updates on things around the farm. My little cousin Josephine, who was ten or eleven, would tell me about what was happening at her school and ask what it felt like to be in the Army. Uncle Paul, who was only two years older than me but not yet drafted, would tell me about which girl he was dating and whether or not he had gotten to touch her breasts yet. Uncle John, Uncle Joe, and Uncle Vincent would send me their versions of what was going on in the war. They would also clip articles and cartoons out of the newspaper, although I think most of what they sent never made it all the way to me, because the Army had censors who went through all the letters, coming and going, for security reasons.

Sometimes Mama and my grandmas would send me a package from home filled with Italian cookies. All the other boys wanted some, and I wanted them to taste and see how good it was to be Italian. Then again, I always saved some cookies just for myself. It was amazing how something as simple as a cookie baked in a mother's oven can comfort a soldier's soul on an otherwise cold and lonely night.

In the evenings, after chow and before lights-out, each of the boys could be seen reading and rereading the day's letters. If I hadn't gotten any mail that day, I would just read a letter from the day before. Some of the guys had girlfriends or even wives back home, so those letters were highly cherished and would be read dozens of times. There were a few boys who wanted to read all their letters out loud. I suppose each of us assumed that the written words of someone we loved would hold just as much significance for the other soldiers. And in a way they did, because we all understood what it was like to be here, together yet alone.

After a while the days started running together. Sometimes it was difficult even to remember what day of the week it was, not that we necessarily needed that information in order to function. But be that as it may, we all rather looked forward to the weekends.

Saturday mornings found us scrubbing the floors and the latrines of the barracks, straightening and cleaning our personal belongings, and then getting suited up in our Class A dress uniforms for weekly inspection. Sergeant Youngblood would pass through the barracks wearing white gloves, wiping the tops of shelves, behind toilets, and under beds for any traces of dust or dirt. If his glove came up anything but white, we would start over, from top to bottom. After the barracks passed inspection, we would stand at attention, shoes polished to a high sheen, uniforms smartly pressed, ties knotted with precision, and posture appropriately military. Sergeant Youngblood had told us at the outset that the only requirement at these gatherings was perfection.

We had Saturday afternoons and all day Sundays to relax and do as we pleased. There was a church in the camp, so most of us went to Mass on Sunday. I suppose they also had Jewish and Protestant services on the weekends, but since I'm Catholic, I only went to the Catholic ones. I generally spent the rest of Sunday resting, writing letters, or just thinking about home. The weekend's respite didn't last long. Monday morning at 4:30 A.M., the cycle began all over again.

March nineteenth of each year is significant for Sicilians because it is the feast day of our patron, St. Joseph. Back home the women would prepare tons of food and place it on a makeshift altar. All of our relatives and friends would come over, and we would say prayers and then eat in honor of the saint. It was always a great day of food, music, and good times. So when March 19, 1943, came around, I was feeling isolated over and above the expected homesickness that all the other boys felt.

At mail call that afternoon, I received a letter from Mama. In it, she told me that Grandpa Sacco—after whom I had been named and who had himself been named Giuseppe as a tribute to the great St. Joseph—had died.

I sat on the edge of my bunk and stared at the words in the letter. I didn't want the other boys to see me, but I couldn't help it, and I started crying. Grandpa Sacco had prayed that I would live through the war and return safely home, and now he had died. I reached into my pocket and pulled out a silver crucifix he'd given me the night before I left home. "You keep this with you, Joe," he had said, "and whenever you're in danger, you hold it in your hand, and you will be safe." Now I held the cross in one hand and the letter in the other. And all I could think was that I would never see him again.

Sergeant Youngblood, who already knew the contents of the letter, approached my bunk. "I'm sorry, soldier," he said as he put his hand on my shoulder. I wiped my eyes with my sleeve as I looked up. "That's okay, son," the sergeant said. "My grandfather passed away a few years ago, too. It's tough. I know it's tough. You gonna be okay?"

"Yeah, Sergeant. I'll be okay. It's just . . ."

"I know. Listen, son, the Army can't let you go home from basic for . . . well, for a funeral. It's not allowed. I'm sorry, Sacco. I really am. But you can go over to the chapel now if you like and say some prayers for your grandfather. You can say good-bye to him." It really grabbed my heart when he said, "You can say good-bye to him." I didn't want to say good-bye—ever.

I walked to the chapel, said some prayers, and thought about Grandpa. After a little while, the priest, Father O'Sullivan, came in and talked to me. It was a good conversation, because I was able to really tell him about Grandpa and how he had come over from Sicily, what he was like, and how much I was going to miss him. I showed him the cross Grandpa had given me before I left.

"You hold on to that cross," Father O'Sullivan said with his Irish accent. "You hold on to it tight. Your grandpa has prayed for your safety and has given you his blessing in the form of that cross. And now he is with God Almighty. And from his perch there in heaven, he will himself be able to watch over you, keep you safe, and return you home to your family when all of this is over."

I turned in earlier than usual that night. The other boys knew what had happened, and they were very respectful. If there's one thing soldiers do take seriously in a war, it's death. The nine o'clock lights-out was a welcome relief from a day that had been the worst of my life so far. Lying in my bunk, all I could think about was Grandpa, my family, and home. They were all together, and here I was, somewhere in the Ozark Mountains freezing my ass off, lonely, tired, unsure of what was happening, and preparing for a war I didn't really want to fight. All I wanted to do was go home.

I could hear some of the other guys snoring. I should have been sleeping by then, but my mind was racing. After several rounds with grief, anger, and self-pity, I was finally overcome by sheer exhaustion and drifted off to sleep.

The next morning, as we stood in ranks in front of the barracks, I felt terrible. I was tired, aching, dizzy, and, as always, cold. The sun wasn't up yet, and a steady wind was blowing sleet directly into my face. During roll call I had the sensation that everything and everyone around me was moving slightly one way and then the other. I had to struggle to keep my balance.

"Sergeant?"

"Yes, soldier? What is it, Sacco?"

"I'm . . . I'm not feeling too good."

"On a beautiful morning like this? I bet you're gonna feel wonderful once we get on our hike."

"Nah, Sarge," I said. "I think I'm sick."

Youngblood walked up to me and studied my face for several seconds. "Sacco, you're all right," he pronounced. As he turned and walked away, everything went black. I barely heard the guy next to me yell, "Hey, Sacco!" And then nothing.

I woke up in the hospital. There were about ten or fifteen other beds in the room with me, all of which were empty except for one at the very end, where a soldier, hooked up to some fluids, was sleeping. A couple of nurses were standing near him talking quietly. I tried to sit up but fell back onto the pillow. One of the nurses, the good-looking one, came walking toward me.

She had dark shoulder-length hair, long legs, and a tiny waist. Her walk was a thing of beauty, her hips moving from one side to the other in the most delightfully feminine way as she approached. Her face was perfect, and she had on just the right amount of makeup. Her eyes were a shade of green that seemed to sparkle when she smiled, and that smile was, in a word, gorgeous.

"Okay, soldier," she said politely, taking the clipboard from the foot of my bed and looking for my name. "Mr. Sacco, is it? Am I pronouncing it right?" She had the most beautiful, fullest lips I had ever seen.

"That's right. You must be Italian."

"No," she said. "I'm not Italian, but my husband is."

"Your husband? Why do all the prettiest girls have to be married?"

"Aren't you sweet?" She smiled as she put her left hand on my forehead. With her other hand, she retrieved a thermometer from her pocket, shook it vigorously, and pointed it in the direction of my mouth. "Now, open up. Say 'ah.' " Her hand on my head was wonderfully soft, and when she leaned over, it was impossible not to notice that she was uncommonly buxom. In fact, I think I got momentarily mesmerized by her oncoming breasts, sort of like a deer caught in the high-beamed headlights of a car. But, unlike the deer, I definitely wanted to get smacked by these things. So as long as they were moving toward me, I was holding my ground. "Come on, Mr. Sacco," she said, smiling. "Are you gonna say 'ah' for me?"

I opened my mouth and took in the instrument, which she placed carefully under my tongue. She leaned back and marked the time on her watch. For a couple of seconds, I was watching her, and she was watching me. I had to speak. "Whtsnuhnneem?" I asked. She looked a little confused, so I repeated myself. "Whtsnuhnneem?"

"I think you'd better wait until that comes out of your mouth before you try to talk," she laughed as she checked her watch again and then walked away. Wow, she looked good walking away, too.

"Nohkeey," I said.

After a couple of minutes, she came back, leaned over me again, and took the thermometer from my mouth. "Hmmmmm," she sighed, studying it carefully. She set it down next to my pillow, picked up the chart, and wrote something.

"What?"

"You have a pretty good little fever there," she said.

"Hey, what's your name?"

"Pardon me?"

"What's your name?"

She realized that's what I'd been trying to ask before. "Well," she said, pointing to the name tag prominently situated high atop her right breast. "I'm Nurse Simmons."

"Oh, I know that," I said. "I mean, what's your name?"

"I just told you."

"Is it Maria?"

"Maria? No," she laughed.

"You look like a Maria."

"Well, it's not Maria."

"Betty?"

"No."

The other nurse—the big, serious-looking one—was approaching.

"Myrtle?"

"No," she said, laughing again as she playfully touched my arm.

The other nurse was upon us. "Nurse Simmons," she said, "are you done here? Because I need you to help me with some paperwork for Dr. Finley. Is he okay?" She picked up my chart and gave it a quick once-over before flipping it back down onto the railing of the bed. "Ummmm," she said. "That's a serious fever. Let's get him some fluids."

"We'll have him back to work in no time," Nurse Simmons said.

"Okay, good," the big one said. "The doctor will be by to see you in a little while, soldier."

The two walked away. I shut my eyes but then heard someone walking toward me. It was Nurse Simmons. "I almost forgot this," she said, retrieving the thermometer beside my pillow. "Are you comfortable?"

"Yes." I wasn't, but I didn't want to complain. Actually, I was thinking, *I'd be a lot more comfortable if you would jump in here with me and keep me warm,* but she was married and all, so I just smiled.

"You need anything?" she asked.

"Nah."

"Okay," she said sweetly. "I'll be back later to check on you."

"Okay."

She started walking away, but when she got to the foot of the bed, she turned and looked back at me with that beautiful smile. "Angela," she whispered, and then left the room.

I was in the hospital for the next two weeks. A few other guys came and went, mostly suffering from injuries sustained during training. Angela and the big nurse both took good care of me. The only problem was that I was too weak to enjoy it. I don't know how I got so sick. The doctor said I was dehydrated and a bunch of other stuff I didn't understand. He told me to rest, drink lots of water and juice, eat all my food, and take my medicine. I really felt terrible, but seeing Angela's face and . . . well, the rest of her made being sick feel good. As long as she was giving medicine, I was taking it.

It was the big nurse who did the muscle work such as helping me walk to the latrine. She's also the one who gave me occasional sponge baths. She'd say, "I'm gonna wash down as far as possible, and then I'm gonna wash up as far as possible, and then you wash possible." She was a nice lady, but I dreamed about Angela performing this particular service. However, it was probably better that she never did. Something tells me that if Angela had touched me with a wet, soapy sponge, the possibilities would have definitely increased, and that would have been embarrassing for the both of us.

Finally I felt good enough to walk, so I got up and went over to the window to look outside. It was snowing. I moved as quickly as I could to get out of the ward before any of the nurses or doctors got back. I went down the stairs and slipped past the girl at the front desk. There was a porch on the front of the building, so I went to one end of it and sat where there was no awning. I was feeling better, but I figured that if I caught a bad cold, I could stay in the hospital longer. In fact, if I got sick enough, they might even send me home. I was wearing one of those thin hospital gowns, and that was all—no socks, no shoes, no hat, nothing but the gown.

I heard the door to the porch open. I recognized those footsteps. "And what are you doing out here, young man?" It was the big nurse. "Get yourself back inside this facility right now!"

I got up and walked past her, trying to breathe in all the cold air I could before I left the porch. But it was too late. I was already well on my way to recovery. I went back to my bed, dejected.

Angela came by to see me later that night. "So I hear you can't wait to get out of here."

"Not really."

"Good news. You're doing much better. Looks like you'll be back on your feet in no time."

"Yeah, that's great."

Two days later the doctor came around and discharged me to my company. He told me that he had given instructions to my sergeant to take it easy on me at first, but that I was as fit as a fiddle and should be at top speed again soon. Angela came by while he was filling out the papers. "You be careful out there, soldier," she said. "We don't want to see you in here again, okay?"

"Okay. Yes, ma'am. I'll do my best." I gave her a little salute.

"This one's a cutie," she said to the doctor before she walked away. As always, she looked great walking away. But, of course, she was married. Besides that, she was a lot older than me—probably twenty-five at least.

My training continued for the next ten weeks. Since the guys I had started with were further along with basic, I was put with another group so that I wouldn't miss anything. Of course, I told all of them about Angela. And, of course, they told me how they would have banged her had they been in my situation. I don't remember—I might have told them I did. Sometimes when you're talking to the guys, you say things to sound tough.

We trained all through the spring and into the first part of summer. The warm weather was a welcome relief, as was the increased daylight. Basic training might not have seemed like it was working, because everybody complained so much, but the discipline and effort did transform us over a period of time. We were healthy, fit, organized, alert, and prepared. Now we were ready to go on to the next level. This is where most of us split up for good, since each man was sent from basic to different parts of the country to join his assigned battalion for maneuvers. I had made some friends through the rigors of training, but now it was time to move on. So, during the second week of June, as my orders prescribed, I boarded a train and shipped out to Fort Polk, Louisiana.

3

War Maneuvers

Fort Polk, Louisiana
June 1943

The weather was getting hot across the South, and Fort Polk, near the town of Leesville, Louisiana, was already starting to swelter. The nearby swamps were producing mosquitoes the size of birds, and the humidity made the air seem thick and heavy. I was more accustomed to the heat of a Southern summer than some of the other guys were, but this was hot even for me. And the worst part of the summer was still ahead.

We were here to practice war maneuvers, which meant we would learn combat and survival techniques by participating in authentically staged battles. Coming out of basic training, this promised to be fun, camping around in the woods, playing war, learning battle strategy, getting shot at but not really getting hurt. In fact, I didn't actually know what we would be doing, but it did feel good to be finished with basic, and I was looking forward to this stage of my training.

When I arrived in Fort Polk, I didn't know a soul. All I knew was that I was to report to the 92nd Signal Battalion. The jeep, which had picked me up at the train station, unloaded me in front of camp headquarters. I hopped out with all my belongings, bounded up the stairs and into the building.

There was a corporal at a desk just inside the entrance. "Can I help you, Private?"

"I'm here to report to the 92nd Signal Battalion"—I had to look at my papers—"Company A."

He reached out his hand, took the papers, and nonchalantly

First Sergeant Ernest Thomas at Fort Polk, Louisiana, June 1943.

glanced at them. He quickly handed them back to me. "Have a seat. First Sergeant Thomas will be with you in a minute."

I don't know how he knew I was there, but within a matter of seconds, the sergeant entered the room. "Private Joe C. Sacco from Birmingham, Alabama?" he asked, extending his hand. I stood as quickly as I could. "I'm First Sergeant Thomas," he continued. "Welcome to Fort Polk."

It didn't take long to realize that this was a man of high energy and intelligence. He seemed a lot older than me, probably around thirty or so, maybe more. Tall, lanky, square-faced, he had the look of a real soldier about him. He wasn't the screaming type we'd encountered in basic; he was more businesslike, more down-to-earth, and much more polite. He was the type of guy you met and immediately liked.

"Okay, Private," he said, "you're in Company A. Let's get you to your quarters." And with that, he was out the door. I picked up my duffel bag and followed. The jeep was still waiting, the driver apparently accustomed to the drill. First Sergeant Thomas jumped in shotgun and I loaded in the back. My butt no sooner hit the seat than we were off.

We drove outside Fort Polk and into the woods. After a couple of minutes, we came to a campsite made up of about a hundred tents. The jeep screeched to a halt. The driver hadn't done this when we drove from the train station, so my thought was that First Sergeant Thomas liked the thrill of the ride. Dust was still swirling around the vehicle as the sergeant hopped out and ran around to the back. "Hey, Private! This is where we get off." I was choking on the dirt kicked up by the jeep, but I jumped out and followed the energetic man up a small path.

"Okay," he said, pointing at a large tent to the left as we continued walking quickly. "This is Battalion Headquarters—HQ. That over there, that's showers. One next to it, latrine. Campsite consists of four companies, A, B, C, and HQ, back there, we just passed it. Each company, three hundred men. Tents arranged by company. Each tent, eight men—ten if you're unlucky. Men in your tent, that's your squad. All the tents look the same, so remember where yours is. Find a landmark—that tree. Tent has a little number, right here. See this number?"

"Yes, 314."

"That's your tent. 314. Welcome home."

The front of the tent was pulled back, creating a door of sorts. Inside, there was a wooden platform floor, presumably so that we wouldn't get muddy every time it rained. There were eight cots, four on either side, complete with a small footlocker and a place to hang uniforms beside each. Netting hung from an overhead beam and cov-

Me at Fort Polk, Louisiana, August 1943.

ered each cot individually. That was actually the first thing that caught my attention, and I looked at it curiously.

"Killer mosquitoes here. They're sons of bitches," said First Sergeant Thomas.

Two GIs were inside. One was sitting on a cot reading a newspaper. The other was lying down on his back, idly tossing a baseball straight up and then catching it, trying to not get it caught in the net. As we entered, they both looked up, but neither one moved or said anything.

"Gee. At ease, men," First Sergeant Thomas said sarcastically. "Don't look so enthusiastic."

"Hi, Sarge," they both said, and then looked at me. I set my duffel bag down.

"This is Joe C. Sacco from Birmingham, Alabama," the sergeant said. "He just completed his basic training. He's part of the 92nd." He pointed to the other guys. "Martin Silverman, New York, New York. Jim Cardini, San Diego, California."

Silverman and Cardini jumped up and shook my hand.

"You know where the shower and latrine are," First Sergeant Thomas said, pointing to his right. "Okay, our mess tent's down this way"—pointing to his left—"all the way down, you'll see it. These two know where it is." He checked his watch. "Okay, it's 1620 right now." We each looked at our watches to confirm the time. "Be at the mess tent for chow at 1800. In the meantime get unpacked and familiar with your surroundings. Find a landmark, remember your tent number, take a walk, get acquainted. Or just read a comic book like Silverman there."

Cardini laughed.

"It's a newspaper," Silverman protested.

"Oh, I'm sorry, Silverman," First Sergeant Thomas laughed. "It looked like the funny-paper section."

"Okay. I like the comics. I have a good sense of humor. What the hell?"

Cardini pressed, "Yeah, Sarge, what the hell? But he has a comic book, don't you, Silverman?"

"Okay. So my nephew sent me a damn comic book. He's trying to cheer me up, keep up my morale. So what?"

"Ah, Cardini," said the first sergeant. "Give him a break. He's just keeping his morale up."

"That's right, Sarge," said Silverman.

"It's First Sergeant Thomas."

"That's right, First Sergeant Thomas."

"You don't even have a nephew," Cardini chided.

Silverman answered, "Yes I do."

"No you don't."

"Yes I do. How the hell do you know what I have?"

Cardini shook his head.

First Sergeant Thomas interrupted. "See, Sacco? See what you've signed up for? Okay, boys, get along with each other. I've got work to do. Sacco, lights-out at 2100, same as basic. Reveille 0400. Any questions?" Before I could say anything, he said, "Good. See you at chow." And he was gone.

I looked at the two guys. "What are your names again?" I asked.

"Jim Cardini, San Diego," said the one with the baseball. He was about my height (six feet), athletic-looking, and somewhere in the vicinity of nineteen years old. He had sandy hair, brown eyes that bugged out slightly when he talked, and a certain swagger that gave me the impression he had not a worry in the world. "You Italian?" he asked.

"Yeah," I answered. "Sicilian."

"No shit?" he said. "A Siciliano. I'm Napolitano. At least my grandparents were."

The other fella, the one reading the funny paper, stuck out his hand. "Marty Silverman, New York." He was thin and a bit shorter than me. He was wearing gold wire-rimmed glasses and had a big nose—not huge, but a little too big for his head. If you looked at him right in the face, he looked like he was maybe in his mid-twenties, but his hairline was receding somewhat, which made him look older.

"Joe Sacco," I said.

"Good to meet you," Silverman said. "I guess we're going to be spending a lot of time together from now on."

"Yeah, I guess so," I said.

Cardini agreed. "Looks that way. Hey, have you met any of the other guys?"

"Nope. Just you two."

"Oh," Cardini said. "We're with some good guys."

"Yeah, they're around," Silverman said.

"All the boys in this tent already here?" I asked.

"A couple still coming," Silverman said. "They should be here sometime today."

"Where can I . . . ?"

"Here you go," said Cardini, pointing to a few of the cots. "Any one of those, your choice."

I chose the one at the end so I wouldn't have guys sleeping on both sides of me. As soon as I put my bag down on the bed, another guy entered the tent. He was a tall, skinny fella with black, curly hair and a jovial smile.

"Hey, hey! How you doing, buddy?" he intoned as he moved toward me, hand extended. "Ed Duthie!"

"I'm Joe Sacco."

"Joe Sacco? Well, I'll be damned. It's good to meet you, buddy. You met Cardini and . . . Silverstein is it?"

"Silverman."

"Yeah, hey, sorry, buddy," he said to Silverman, and then turned back to me. "I just got here. Still learning everybody's name. Your name is S . . ."

"Sacco."

"Okay, that's easy. How's it spelled?"

"S-a-c-c-o."

Duthie raised his eyebrows and flashed a big smile. "Just the way it sounds. Sacco!"

About then another guy poked his head inside the tent. This guy looked vaguely familiar, with his green eyes and wiry frame. "Hey, guys, how's it hanging in here?"

Cardini said, "A little to the left, and it's making my right ball jealous."

"Real good," Duthie said. "We got a new man."

The guy at the entrance stared at me curiously. "Hey, I know you," he said.

"You do?"

"Yeah. Where were you in basic?"

"Camp Crowder. I got here just now."

"Hell yes." His face lit up. "That's where I know you from! From the truck. I met you the first day I was there. Birmingham, Alabama, right?"

Now I remembered who he was. "That's right."

"Paul Averitt. Nashville, Tennessee."

"Yeah," I said, approaching him to shake hands. "Hey, I never saw you after that. Whatever happened to you?"

"Oh, hell," Averitt answered, shrugging his shoulders. "They shipped my ass out to Texas after a couple of days. I went all through basic there. Must be the Army's way of being efficient. Who knows? So what are you doing here? You part of the 92nd?"

"Yep."

"No kidding?"

"News flash," Silverman said. "We're all in the 92nd Signal Battalion."

Averitt laughed. "You met Silverman?"

"Yeah," I said.

"Gets a lot of his information from the funny paper."

"Geesh," said Silverman. "So I happen to look at the damn comics now and then."

Another guy was entering the tent. "Hey, Averitt, what are you doing in here with the men? I thought you lived next door."

"I'm greeting our new brother in arms," he said, pointing to me. "Joe Sacco, from Alabama."

The soldier looked at me. He was a short fella with a ruddy complexion, high cheekbones, jet-black hair, and the whitest teeth I had ever seen. He spoke with a slight accent of some sort and didn't look American, but I couldn't figure what country he was from. "Hey," he said, coming in my direction. "Charles Spotted Bear. Midland, Texas." He shook my hand vigorously and smiled a toothy grin. "Hey, see this guy back here?" he asked, whipping around and pointing to a chubby little soldier standing behind Averitt. "That's Sam Martin."

Sam Martin waved without saying a word. Averitt looked around. "Son of a bitch! How the hell did you get right up behind me without making a sound?"

Sam Martin smiled and said, "Our people can always sneak up on the paleface."

"Go to hell," Averitt laughed.

"It's true, Averitt," Spotted Bear said. "We will educate you in our ways. You'll see." Averitt scratched his forehead with his middle finger. Spotted Bear turned back to me. "Sam Martin and I are Indians."

"Indians?" I asked.

"Indians."

"As in 'cowboys and Indians,' " smarted Averitt. "But we like 'em anyway."

Spotted Bear was a good-natured kid, and he started laughing. "Yeah," he said, "as in '*Indians* and cowboys.' When we used to play, the Indians always won." Sam Martin was in the back looking serious.

I thought this was the neatest thing in the world. I had seen Indians in the movies, but I had never met one in real life. "Your name is Spotted Bear?" I asked.

"Yeah, that's right."

"So . . . you named after a bear?"

"Yes. A spotted one. You named after a sack?"

"I'm Italian."

"Italian?" The Indian was studying my face.

"Sicilian," Cardini piped up.

"Like Al Capone?" Spotted Bear asked.

By now another guy had poked his head into the tent. He was about five foot nine, with dark, wavy hair, blue eyes, and a friendly air of confidence about him. "What's going on?"

"New guy," said Averitt.

"Joe Sacco," I said, moving to shake hands.

"Hey. Jim Hodges. Sacramento. You just get in?"

"Yep."

"You know any of these guys from before?"

"Nope."

Averitt said, "We met at basic for one minute."

Hodges looked at me. "Maxey?"

"Camp Crowder," I answered.

"Okay." He turned his attention back to Averitt. "Hey, Averitt, the first sergeant said to tell you he wants you to come by HQ before chow to pick up some papers."

"Yeah, he already told me. I just saw him when I was walking up here. I'm heading down there in a couple of minutes."

"You in trouble already?" Hodges asked.

"Nah," Averitt responded. "Just mind your own business."

I got Cardini's attention. "Y'all know each other before now?"

"*Y'all?* Hey, listen to this," he laughed.

Spotted Bear answered my question. "A few of 'em were in basic together. Me and Sam Martin been here for two days. Cardini was here when we got here. Silverman, how long you been here?"

"Since yesterday afternoon. But I was in basic with Averitt."

"And I just got here this morning," added Duthie.

Spotted Bear looked around. "I think there's a couple more coming. They should show up today."

"Where you from, Duthie?" Cardini asked.

"Tacoma, Washington."

"Tacoma, Washington?" Silverman echoed.

Spotted Bear looked confused. "Where the hell is that?"

"Oh, it's up on Puget Sound, just south of Seattle," Duthie answered.

"That's part of the United States?" Spotted Bear asked.

Duthie said, "Oh, yeah, it's part of the United States. It's up near—" He started trying to draw a map in midair with his arms.

"That's okay," Spotted Bear interrupted. "I believe you."

We all had a good time horsing around and poking fun at each other; it was our way of loosening up, of getting to know each other better. After being with these boys for fifteen minutes, I felt at ease. All the rest of the guys showed up within the next day or two. They seemed like good, decent fellas. Most of them were nineteen or twenty. Some were older. At eighteen, Duthie, Averitt, and I were the youngest.

All of us understood that we would probably be together for quite some time from here on out, so we made an extra effort to get along. Even in basic the Army had drilled it into our heads that we were responsible for the safety of our buddies. They had told us that we literally had to trust each other with our lives and that if we couldn't do that, we'd be flirting with disaster from day one. There were a lot of different personalities, but I was glad to be with these men.

Of course, in any group, there's always the asshole or two. And an animal as large as the U.S. Army has many, many assholes. So you get to know who they are over a period of time, and you keep away from them. I don't know if it was because I was the youngest of the group or because I never went looking for trouble, but most of the asshole guys left me alone. Of course, sometimes you can't keep away, especially if he's in the same tent.

In this case his name was Chicago. At least that's what he called himself. Some of the guys would nickname themselves the city they were from. I couldn't imagine anyone really named "Chicago," but then again I wouldn't have ever imagined anyone named "Spotted Bear" either. Anyway, this guy Chicago was sort of a hothead, always ready for a fight. In fact, he had a curiously flat nose and a scar just above his right eye—indications that he'd taken a few more punches than he'd dished out. He was slightly smaller than me, but had a brooding disposition that seemed to keep him constantly agitated. He enjoyed making fun of Silverman, probably because Silverman was a Jew. But he also irritated the hell out of Duthie, who was the most easygoing guy we had. I think he just liked to see if he could get people riled up.

One day we were sitting around the tent talking, and I must have said something he didn't like, and he said, "Sacco, you think you're tough, don't you?"

I said, "No, I don't."

He said, "I can whip your ass."

"I'm not interested in fighting you."

"Afraid I'll whip your ass, huh, Sacco?"

So I said, "Hey, guess what? I'll be happy to knock the hell out of you right here and now, and we'll see who's tough." I really didn't want to fight him, because he probably would have kicked the crap out of me, but I couldn't back down.

Spotted Bear looked up from the letter he was reading, stood, casually strolled over to Chicago, and said, "I'll tell you what, hotshot. If you touch Sacco, I'm gonna come to you in the middle of the night and scalp you. I know how to do that. Me and Sam Martin do that all the time to people who piss us off back home. I would love nothing more than to see them ship your body back home without a head." Chicago's eyes got wide. Spotted Bear knew he'd gotten his attention. "And they won't even put me in jail, because I'll tell them I scalped you for religious reasons. I'll tell them I had to get rid of the evil spirits." He made a scalping motion with his hand. Chicago swallowed. The Indian started to walk away, then turned around. "And another thing: Stop picking on Silverman."

Spotted Bear made that motion with his hand again, went back to his bunk, picked up his letter, and continued reading. After that,

Chicago was real nice to us. A few days later, while we were walking to chow, I asked Spotted Bear, "Hey, you really scalped people?"

He gave me a wide smile, "Hell no. But there's a first time for everything, my friend."

Our training at Fort Polk included some technical aspects of the signal corps as well as military logistics and maneuvers. We already knew how to handle a weapon, march, carry everything we owned in a backpack, and go for days eating nothing but C rations. We had learned that in basic. Now we were being trained how to do all of that while setting up communications for the infantry and not getting killed at the same time. In order to do so, we had to become proficient in the rudiments of stringing telephone cable high in trees, through the forest, over hills, and across rivers, all the while engaged in mock battles and skirmishes.

We'd get up at 4:00 A.M. and fall into ranks in the yard area beside the tents where First Sergeant Thomas would call the roll. Then we would exercise for about half an hour. It was usually quite warm even before the sun came up, so within a few minutes, we'd all be sweating like pigs. On some days, the sergeant would have us take our shirts off and lay them on the ground in front of us. Then we'd run through the woods for a mile or two, come back, and begin our exercises. After that, we'd put our shirts on and go over to the mess tent for breakfast.

I still loved a good breakfast, so I didn't mind doing the running and the exercise so long as I could return to the mess tent and pull up a big stack of pancakes and some eggs and sausage. I usually drank a couple of glasses of milk, but eventually I started getting in the habit of drinking coffee, or "joe," as the guys called it. This took some conditioning, because I had never really been a big coffee drinker. At first I just liked the way it smelled when the other boys drank it. So I started drinking it myself. Before long I was addicted. The mess sergeant made it so strong that it could bring the dead back to life. If all the running and exercise didn't wake me up in the morning, the coffee sure enough did.

After breakfast we'd take a quick shower. We had, on average, approximately fifteen minutes during which we were expected to shower, shave, use the toilet, and get dressed in our Class A or Class C uniforms, depending on our duties that day. Class A was for dress purposes only, so we were usually in the Class C fatigues, ready for work.

The shower tent was a wooden platform on top of which were pipes with shower spigots attached. The tent was cordoned off in sections by pieces of canvas about shoulder high, so we had some semblance of privacy. The water was usually hot—for the first thirty seconds. Then we'd get a case of the blue balls. That's when the canvas came in really handy, because nobody wanted to look too shriveled up in front of the other guys.

Hot, cold, or otherwise, whatever we did in the morning, we did it within fifteen minutes. Then we were dressed and out front for First Sergeant Thomas to give us our orders for the day. Sometimes we'd jump into a truck that would bring us up to the front lines. Other times, we'd just hike it. Either way, we'd get back around 1700 or 1800 (5:00 or 6:00 P.M.), eat, take that optional second shower, which felt good after a day of sweating in the field, and then relax in the tent until lights-out at 2100 (9:00 P.M.).

There were times, of course, when we'd just stay out in the field for days at a stretch as part of the war maneuvers. On one of our first overnight maneuvers, I was teamed up with a guy named Rodgers from my tent. He was originally from someplace in Michigan. We each carried our own supplies in our backpacks, but since we were a team, he carried half of a pup tent and I carried the other half. When it got dark, First Sergeant Thomas told us to pitch our tents in the woods for the night. The battalion was pretty spread out, because this is the way it can happen in a real war, so each tent was to have a guard on duty all night.

I wanted to pitch our tent a little deeper into the woods, but Rodgers insisted on setting it up beside a tree right next to the road. He was older than me, so I went along.

"You take first watch, Sacco," he said.

"Okay. No problem."

"Great." He looked around, retrieved a couple of things from his backpack, got up, and started to walk off.

"Hey, where you going?"

"I'll be back. Stop worrying," he said. "I'm just going into Leesville to see my girlfriend. I'll be back by my watch."

"Like hell," I said.

"'Like hell' what?"

"Like hell you're going to see some girl. You stay here like the rest of us!"

"Oh, kiss my ass, Sacco. I'll be back in plenty time. You just stand your watch and you make believe I'm in there sleeping." And with that he was gone.

The signal battalion was neutral in the maneuvers, since we supplied communications for the umpires and both sides, or teams, as they waged their wars. Whereas the teams wore red or green armbands, we wore white ones. And whereas they maneuvered out of sight of each other, we were out in the open, stringing telephone wires for each of the front lines back to their respective headquarters. So Rodgers would have easily been able to slip away without getting captured by either side. They would just think he was on some signal battalion business and let him go on his merry way.

As the neutral team, we had no lights-out policy during these mock battles. I therefore built a campfire in front of the tent and stood my watch. All four hours of my watch went by, and no Rodgers. I watched some more. No Rodgers. Thirty minutes. One hour. Two hours. After a full day of maneuvers, I was getting real sleepy and could hardly hold my eyes open. Then it started to rain. *Damn,* I thought. *I'm not staying out here in the rain all night so that that asshole can be sleeping in a comfortable bed with his girlfriend. He's gonna show up at four o'clock. Screw him.* So I put out the fire, climbed inside the tent, and went to sleep.

Early the next morning, I was awakened by a rumbling noise that got louder and louder. I could tell that it was getting light, but I didn't want to get up. Suddenly my tent was gone, and I was staring straight up into the tread of a tank, which came to a loud, creaky stop just above my nose. I was more than a little confused. I slid down a bit and stood up. What I saw shocked me. Apparently a Sherman tank had slid off the muddy road and wiped out my tent. The thirty-ton vehicle had moved to within a couple inches of my head before it miraculously stopped.

The tank commander jumped out. "Oh, shit!" he yelled. "Oh, shit! You okay?"

"I'm okay. What the hell were you doing? You working for the Nazis? You trying to kill us before we can get over there?"

He looked up under the tank. "You alone?"

"You see somebody up under that tank?"

"No."

"Yeah, I'm alone."

"You're supposed to be in pairs."

"Other guy took off," I said.

"Took off?"

"Went to see a girl in Leesville."

"What? You're supposed to have someone standing watch while you sleep. I didn't even see you here. I could've killed your ass! Stay here," he ordered—as if I were going somewhere. He climbed quickly back inside the tank. I rooted around until I found my backpack. He must have radioed to someone, because within a few minutes, First Sergeant Thomas was barreling down the same muddy road in a jeep. The driver hit the brakes and slid sideways to a stop. That's the way the first sergeant liked it.

He hopped out, looked at the tank and what was left of the tent, and then came over to me. "Sacco, you okay?

"Yes, First Sergeant."

"What happened here?"

I told him what I knew.

"Your partner is supposed to be on guard in order to keep *that* tank away from *that* tent," he said, pointing at each in dramatic fashion. "Okay, get your backpack and get in the jeep."

He drove me down the road a bit and teamed me with Duthie and Spotted Bear for the rest of the maneuver. I never saw Rodgers again. I never knew what they did with him. They had his area all cleaned out by the time we got back to camp a couple of days later.

Since we had an extra bunk and since they had ten guys in 315, First Sergeant Thomas moved Averitt in with us. Our tent, 314, therefore consisted of the following men: Averitt, Duthie, Spotted Bear, Silverman, Cardini, Chicago, myself, and a guy named Chandler.

Tom Chandler was about thirty years old, at least. He seemed older. He was a big, rough-looking man, completely bald, with a couple of scars on his chest and a tattoo of a cross on his left arm. The story was that he was an ex-con from Joliet Prison. None of us really knew what he'd done to get there, although we each had our own theories. Apparently they had given him the option of staying in prison or joining the Army. And now he was in our tent.

He was known for doing crazy stuff like taking a shovel and smashing an Army jeep with it. He spent a few days in the Fort Polk stockade for that. He'd go AWOL, go into Leesville, and bust up a bar without a second thought. The military police would go get him out of the local civilian jail and lock him up on camp. Why he didn't get sent away like Rodgers, I'm not sure. Most of the guys kept their distance from him for fear that he would go crazy on them. He was pretty quiet around the tent—sometimes too quiet. He would sit there and look at the paper for a long time. We figured he was reading about some crime he'd committed or checking to see if they'd discovered a body he'd buried in a shallow grave. Sometimes he'd just look up and stare at Chicago. I thought that was funny. Chicago didn't.

Chandler was one of those guys who would go to sleep within one second of lying down. We knew this for a fact, because he'd be snoring so loud that people on the other side of Fort Polk could hear him. It sounded like we had a bear in the tent with us. Sometimes at night, while we were all sleeping, he would just scream out loud, then turn over and go back to sleep like nothing ever happened. That prompted lots of theories.

Chandler tossed a table across the mess tent one day because he found a bug in his mashed potatoes. I suppose he was accustomed to all that five-star dining at Joliet. Anyway, all hell broke loose. The only one who got in trouble was—of course—Chandler. They put him in a room while they waited for a jeep to bring him to jail. First Sergeant Thomas pulled me aside.

"Sacco, you guard Chandler until the MPs transfer him to the stockade."

"Hell no, First Sergeant! I don't want him to beat the shit out of me!" (Chandler had a habit of beating up the MPs who were trying to guard him.)

The next thing I knew, I was sitting in this little room with Chandler. I was staring at him. He was staring at me. I looked at my watch. I looked out the window. I looked at my watch again. I looked up at Chandler.

"Aw, hell, Sacco," he said. "Quit worrying."

"Whataya mean?" I tried to act like I didn't know what he was talking about.

"I'm not gonna escape. I beat the hell out of those MPs because they're assholes, but I don't wanna get you in trouble. You look too innocent. Just relax."

When the MPs got there, they found Chandler and me sitting in the room talking quietly about what it was like when we were boys. As they were taking him out, I said, "Thanks, Chandler."

He said, "Okay," and then smiled. I think he beat the hell out of one or two of them on the way to the stockade. He wasn't around the tent for the next couple of weeks.

Maybe hanging out with Chandler wasn't such a good influence on me, because a few days later I got in trouble myself. Here's what happened:

The weather was hotter than hell. In some parts of the country, it cools down at night, but not in the South. Here it just gets dark and stays hot. They say it's because of the humidity holding the heat in the air. Whatever it is, it's miserable, and when the night is breezeless, it gets difficult to breathe and almost impossible to sleep. And you can double that if you're in a tent filled with snoring, tossing and turning men.

Those mosquito nets over our cots seemed more effective at keeping out fresh air than mosquitoes and flies, which took great joy in dive-bombing our heads all night. Some nights I would just sleep through it all with no problem. Other nights I would lie awake, distracted by the sweat rolling down my face, the growing heat of my pillow, and the unending sorties of the insect air force.

One late July night, it was particularly hot, and I was having trouble getting to sleep. There was no breeze whatsoever, and the humidity made it feel like we were underwater. Under hot water. There was a constant and irritating buzzing sound everywhere and the resulting slapping noises when the creatures would land on one of the guys. Each slap would in turn be accompanied by a "Shit!" or "Bastard!" or worse. But there was little else for us to do, other than swat and cuss. To make matters worse, a cricket got under our floorboard and was chirping as loud as it could. All the guys were restless, and somebody across the way—I think it must have been Cardini, because Chandler wasn't there—kept farting about every fifteen or twenty minutes. By the way, a fart is the last thing you need in a tent on a hot, breezeless night.

I wanted to throw something in the general direction of the farter, but what little energy I had was spent swatting mosquitoes off my face.

Finally, after hours of suffering, it cooled down enough for me to fall asleep. Maybe the mosquitoes had been called back to their head-quarters to prepare for their morning raids against us, or maybe I just passed out and couldn't feel them slamming into my head anymore. Either way, I was out. Unfortunately, it didn't last long. The dreaded bugle soon sounded. It was 4:00 A.M.

First Sergeant Thomas whacked on the post of our tent and blew a whistle as he passed. "Time to rise and shine, ladies! Rise and shine! We've got work to do!" He continued on up through the rows of tents.

I put my pillow over my head. "Shit."

Most of the boys were pulling themselves out of their cots, shuf-fling around for their socks and pants. Sergeant was coming down past our tent again. "Let's move it, time to get up and move. Let's go. Roll call, fifteen minutes." He whacked our tent once more and passed on by.

I looked around and saw Averitt lying in the next cot with one eye partially closed and the other staring straight up like a zombie. Duthie was a couple of cots over lying on his stomach, his curly-haired head buried under his pillow.

Averitt turned his open eye slightly toward me. "I say screw 'em," he said.

"Huh?"

"Screw 'em," Averitt repeated. "I'm not getting up."

"Hey, Duthie, you hear that?" I asked. There was no response. "Hey, Duthie?"

"Shut the hell up, Sacco," Duthie mumbled from beneath the pil-low. "Yeah, I heard him, and I agree. Good night."

"Come on," I said. There was no motion. "But . . . ," I began. I was too tired to continue.

"Screw 'em, Sacco." Averitt was getting annoyed. "You want to get up, then get up, damn it. I'm going to sleep. They can't do anything to you. Screw 'em."

Duthie made a noise, quite unintelligible, and then began snoring. Averitt put his arm over his eyes and got quiet. I turned onto my side and, despite the racket of the other men leaving the tent, drifted into a sound sleep. Within minutes a stick resonated off the foot of Averitt's cot. He opened both eyes and slightly lifted his arm until he could see the outline of a rather irritated First Sergeant Thomas.

"What the hell's going on here, soldiers?" Thomas boomed.

Averitt muttered something that sounded like, or at least rhymed with, "Go to hell."

"What did you say to me, soldier?"

I lay as still as I could.

"I'm trying to sleep," Averitt said.

"So," the sergeant began, "you boys are trying to sleep. Well, I hope to hell the rest of us are not disturbing you."

"You are," Averitt interrupted.

First Sergeant continued, "Sacco, Duthie, you trying to sleep, too? You boys need your beauty rest as much as Averitt here?"

Duthie made a grunt. I just tried to act like I couldn't hear anything. Finally the sergeant turned and walked away. "I'll tell you what," he said before he left. "When you girls decide to wake up, go on over and see the captain."

I wished I'd gotten up and gotten the hell out of there when I was supposed to, but I hadn't. I figured I was already in trouble at this point, so I just went back to sleep. Screw 'em.

We woke up about four hours later, got dressed, and headed to HQ for a meeting with a very unsympathetic captain.

The three of us spent the next two days digging holes in a field next to the training area, filling them in, and then digging bigger ones beside them. It was twice as hot as hell and ten times more humid. The giant mosquitoes and gnats were overjoyed to see three shirtless, shoveling, sweating men on whom they could feast, and they were quick to notify all their flying friends and relatives to come by and have a bite or two.

These had to be the most miserable two days of my military life so far. The experience was meant to be a punishment, and it was. It was meant to teach us a lesson, and it did. We had willfully disobeyed orders, and now were paying the price. Even so, as we shoveled, Duthie and I took solace in constantly blaming the whole episode on Averitt, who in turn constantly told us to go to hell, to eat shit, to drop dead, and any number of other things that didn't seem quite as bad as what we were doing. One thing was for sure: The next time First Sergeant Thomas said to get up, I was getting up.

As the dog days of 1943 passed, we settled into a less-than-relaxing routine. Up early, exercise, hike, string cable, eat, work, hike some more, sweat like a pig, ride in the truck, string more cable, set up a phone line to HQ, listen to bullshit from the other guys, and hope you get some mail—all the while swatting flies and gnats in the blistering heat of the Louisiana summer. We wore our Army fatigues, which translates to "hot," and usually carried a full backpack, which translates to "heavy." We had been issued a plastic helmet and were required to wear it during the maneuvers. Wearing a helmet, even a plastic one, is not like wearing a sun hat. In fact, it's hotter than hell. It would basically catch the afternoon heat rising from the ground and concentrate it around our heads. But, like everything else, we got used to it, because we didn't have a choice. Before long we realized that in addition to being protective headgear, the helmet could also serve as a chair, a pillow, and, with the addition of a little soap and water, a washtub.

Speaking of water, it seemed as if every afternoon brought its own torrential downpour, replete with lightning, thunder, and the associated rise in both humidity and bug life. The gnats, mosquitoes, and flies were, for lack of a better term, pains in the ass. And the afternoon rains seemed to whip them into a feeding frenzy, causing them to swarm around and aggravate the hell out of us. It was impossible to eat without them getting all over our food. Chandler would become furious and start cussing the airborne critters. Spotted Bear and Sam Martin claimed they knew some magical Indian way to keep the insects off them, and the funny part was, they never seemed to have any mosquito bites. Chicago said it was because the mosquitoes didn't want their impure, nonhuman Indian blood. Spotted Bear threatened once again to kill Chicago in his sleep, and the wisecracks stopped.

All in all, we just went about our business, doing our jobs. We were trying to learn everything we could, trying to get along with each other, trying to do the best we could under the circumstances. It's not like we were having a great time. In fact, if the truth be told, I'm sure all of us wished we could just go home and forget the Army.

There were some tragic moments when, I suppose, the pressure got to be too much for some of the boys. One infantryman involved in the war games on the green team killed a second lieutenant during the maneuvers. They said that the lieutenant had been getting on

this soldier's ass about something. The guy snapped and put a bayonet through the lieutenant, stuck him to a tree, and just left him
hanging there until some of the other infantrymen found him a few
hours later. After that, we all tried to make sure Chandler never got
too upset.

On Saturdays, after inspection, we were allowed to relax, rest, and
do whatever we liked. Some weekends we would get passes and go
into Leesville to have a little fun. Of course, whenever we left Army
property, we were required to wear our Class A dress uniforms and to
conduct ourselves in a professional manner. Sometimes boys from
one of the other outfits would get drunk and get into fights at the
local bars, but we steered clear of that, because we didn't want to get
picked up by the MPs and get in trouble with First Sergeant Thomas.

Once, while we were getting on the bus to go into town, Chandler
came up to me and said, "Hey, Sacco, come back here and sit with me
for a minute." All the rest of the guys were up near the front. There
were some civilians on the bus, and Chandler and I were almost at
the very back. He waited until everybody was involved in conversation and then pulled a fistful of letters from his pocket. He handed
them to me.

"What's this?"

"Letters from people," he answered.

I was confused. "Yeah, I can see that. So why are you giving them
to me?"

He was whispering. "Because I want you to read them."

"I don't want to read your damn mail."

"No," he said quietly. "I want you to read them to me."

"What?"

"Read these letters to me." There was a silence as I studied his face.
"I . . . I can't read. I need somebody to read them to me," he said.

I was amazed. When I first went to school, I couldn't even speak
English, and they taught me how to read. I was just staring at Chandler, looking at the expression on his face, and I started to feel sorry
for him.

"But you sit there and read the paper all the time," I said.

"I . . . I don't really know how to read it. I just want to make the
guys think I can read it, so I look at the pictures and . . ."

I wanted to ask him a lot of questions, like, "How the hell can you not know how to read?" and "Didn't you have to take a test to get into the Army?" and "Didn't you go to school?" But instead of asking anything, I just took the stack of letters from his hand, opened the first one, and started reading it quietly to him. As soon as I started, he stopped me. "Hey, Sacco."

"What?" I looked up.

"Don't tell any of the other guys . . . uh . . . about this, okay?"

"Okay, Chandler. Okay." I looked back down at the letter and began again. " 'Dear Tom, All of us send our love, and we miss you very much. Becky got a—' "

"Becky's my sister," he clarified.

"Okay. 'Becky got a job at Mr. Whitman's hardware store. . . .' " I read Chandler his letters all the way to and from Leesville that day. He also wanted me to write back to his mom and his sister. I didn't mind. The one thing we all wanted to get was letters from home. I kind of felt bad for Chandler because he couldn't enjoy the letters on his own. But once I would read them to him, he would look at them, page by page, as if he were reading every word. I think it just made him feel good knowing what was in them.

Every day, just after mail call, the guys would go off by themselves, each reading his own mail, each quiet, with his mind a thousand miles away. It felt great to get a letter, but in a strange way it also made me feel rather lonesome, because while it brought someone's words right into my hands, it also reminded me that I wasn't home. And despite all our training and activities, there was still plenty of time to be homesick. Sometimes I'd see a guy get real down because he was thinking about home. We'd say he "got the red ass." I don't know where that saying came from, but that's what we'd say. I know we all got it.

The good news was that maneuvers for our battalion were scheduled to conclude on September 12, and First Sergeant Thomas had told us that we would each be given a two-week furlough before we had to report to the next phase of our training. I hadn't been home since February. It seemed like forever. Like all the boys, I couldn't wait to get home, see everybody, and revisit the way it used to be.

**Me at home on furlough in
Birmingham, Alabama,
September 1943.**

**Birmingham, Alabama
September 13, 1943**

Papa, Uncle Vincent, Uncle Paul, and several of the younger cousins were waiting at the Birmingham train station to welcome me home. We were, of course, required to travel in our Class A uniforms, so my cousins wanted to salute me. I was laughing, because nobody salutes a private, except for maybe a kid.

Mama had cooked up a big feast. I remember that everybody came over that night to see me and to ask questions about life in the Army. It felt so good to be home again with all the people I loved. I had plenty of stories to tell them, but they also had stories to tell me—just little things that had happened around the farm since I'd been away. I was less concerned with talking and more concerned with listening, because my heart had always been at home, and I wanted to know about everything I'd missed.

It was difficult not having Grandpa Sacco around. Everybody else seemed to have come to terms with his being gone, but I kept expecting him to walk around the corner any minute. I could tell that Grandma still really missed him. The second day I was home, we all went out to the cemetery and visited his grave. I'm sure the others could see that I was having a hard time with it. I pulled the cross from my pocket—the one he'd given me—and touched it to the photograph on his headstone. That made me feel a little better. There's a saying that funerals are for the living, and I was beginning to understand what that meant. Not having had the chance to say good-bye had made it so hard for me to accept Grandpa's death, but by the time we left the cemetery, I felt at peace.

The rest of my furlough was spent on or around the farm—and that's exactly where I wanted to be. Some days I helped with the chores. Others I just wandered around the farm, remembering feelings of what seemed like so long ago. In the evenings the women prepared food and we ate, once again beneath that giant tree beside the house. The laughter and stories at the dinner table were now more cherished than ever. The food tasted even better than I had remembered and the wine even sweeter. Nothing the Army cooked up could ever compare to this.

Me with my mom, Rosa, in Birmingham, Alabama, September 1943.

The warm evening breezes coming over the fields carried the unmistakably fresh aroma of Italian herbs and mint. The smell was as familiar as my past, yet I don't think I had ever really noticed it before. And even when I'd tried to recall every detail of this place, I'm sure I never remembered hearing the beautiful music of the birds as the sun was setting. How could I not have noticed such a sound? How could I have taken it for granted? For me the beauty of this visit was that I was given another chance to soak in all the details and love of this place called home.

The night before I left, all the cousins, relatives, and friends came over to say good-bye. This time we all knew I would be gone for a long while, maybe forever. Everybody had something nice to say. A couple of the older folks took the opportunity to give me advice. They didn't know anything about our training, so I just let them talk. One man, Mr. Tony Sciatta, pulled me to the side and showed me a small cloth sack with two acorns in it. "You keepa thisa with you alla the time, Joe," he told me in whispered tones as he shoved the talisman into my coat pocket. "Whenever you needa gooda luck, you reacha inna you pocket anda grabba these nuts." I smiled and said thanks, but I was thinking, *Yeah, if I'm scared, I'm gonna grab some nuts all right, but it ain't gonna be these.*

Leaving home was never easy, but at least this time I knew what to expect.

Communications Training

I stepped off the train at Camp Maxey, near the town of Paris, Texas, with the usual reluctance of a soldier returning from furlough, but also with the added confidence of a man with an extra set of nuts in his pocket. I wasn't the only one who arrived with a little something extra. A couple of the boys in Company A were said to have contracted a case of the clap during their time home. On a much more uplifting note, I soon found out that Duthie had come back with something he could really be proud of.

Averitt was lying on a bunk when I entered our barracks for the first time. The other boys were just starting to arrive. After living in a tent for the summer, the look and feel of an otherwise austere Army barracks was downright homey.

"Hey, Sacco! How's it going?" Averitt asked.

"Good. So you made it back?"

"Just got here this morning. Here, take this bunk next to mine."

I said, "Nah, I'm going down here. I want to be on the end." I also didn't want Averitt talking to me all night, trying to make me laugh.

"Hey, that's a good idea. I'm moving down there, too," Averitt said as he packed his things and followed me. "Hey, you hear about Duthie?"

"Duthie? No. What about him?"

"Got married. Don't tell him I told you."

Before long, Duthie bounded into the building, all happy, all smiles, greeting everyone along the way. "Hey, guys! You down there?

I'm moving down there, too." Duthie picked up his duffel bag, which had been leaning against a wall, and hurried toward Averitt and me. "Hey, Sacco, did I ever show you a picture of my girlfriend, Joan?"

"Yeah, about a thousand times."

"Okay," he said as he pulled the same photograph from his shirt pocket. "Now I want to show you a picture of my *wife!*"

"Your wife?" I acted as surprised as I could. "Hell, congratulations, buddy!" I shook his hand vigorously. "That's great! Let me see that picture again." I studied the photo for a moment. "I can't believe such a pretty girl would marry something as ugly as you!"

"But he looks sharp in his uniform, don't he?" Averitt added.

Duthie took the picture from my hand and gazed lovingly at it.

"Duthie," I blurted out in a moment of uncontrolled expression, "you must be crazy!"

Duthie was shocked. "Why do you say that?"

I had started, so I was obliged to finish. "What if you go over there and get your ass shot? What are you gonna do? Leave that poor girl a widow?"

"I'm not planning on getting shot. I'll be back," he said, smiling and looking at the photograph. "I'll be back." He glanced up at Averitt and me. "All three of us will be back." He scanned the room. "We're all coming back. . . ." His voice trailed off, and he stared down at the image of his new wife.

All that mattered was that he was happy. It made me feel good to know that he was able to capture a piece of home, in a sense, and tuck it safely away in his heart before we shipped out. I liked the fact that he'd made the commitment that he would return in one piece and that, for him, love was stronger than anything we were about to experience. We were still getting to know each other, but I already knew that Duthie was a good man, and I was genuinely happy for him.

It was interesting to me that out of the forty or so men in our barracks, all of us from tent 314 ended up bunking at one end of the room. I suppose we'd spent so much time together during maneuvers that we felt like we belonged together. But, in a way, it was good to hang out with some of the other guys on a more regular basis.

Sam Martin was in our barracks, which made Spotted Bear happy, because they'd been buddies back home, plus they liked to talk about how they were Indians and all. There was a big, burly guy named

McCoy who always seemed to be in a bad mood and who insisted on being called "Tex" even though he was from Montana. There was the ever-smiling Jim Hodges from California—he's the one who had popped his head into our tent the first day of maneuvers. Hodges, as opposed to Tex, seemed perpetually calm, relaxed, and assured that everything would turn out fine. There was Gene Myers from some-place in Illinois and Johnny Vincent from New Orleans. There was MacDonald, Jackson, Serio, Wilkins, Fox—the list went on and on. It was difficult to remember all the names at first, but we were around each other so much that it didn't take long to figure out who was who.

All of us were here in the far reaches of Texas to master the required technical aspects associated with our duties in the 92nd Signal Battalion. Basically, we had to become proficient at providing constant and reliable communications between all parts of an army, from an advancing infantry to a general's headquarters miles away and all points in between. This meant we had a slightly different regimen than we had experienced before, most notably the attendance of after-noon classes and daily field exercises in combat communications.

First Sergeant Thomas was still with us. He told us he would be with us through the whole war, which made all of us feel good. He was strict but fair, and he always had a way of maintaining discipline. He was a tough soldier who had been in the infantry, so I suppose he must have thought that guys in a signal battalion were a bunch of goof-offs. He did get irritated with us, especially at first—and usually for good reason—but as time went on, he mellowed a bit, and we gave him fewer reasons to get upset. All in all, he was pretty patient. Besides that, we were the type of guys that had a way of growing on you.

Our day began at the relatively late hour of 5:00 A.M. with exercises and calisthenics. Some mornings we would put on our Class A dress uniforms, go over to the parade ground, and march as the Army band played John Philip Sousa music. First Sergeant had said that when we march in review, we should have precision in our gait and resolve in our attitude. I for one loved these mornings. There was something about hearing that music and seeing thousands of soldiers marching with one common goal—and being part of it all—that was inspiring to a farm boy from Alabama. I think the Army knew that.

First Sergeant Thomas had taught us how to do different parade maneuvers with our rifles, such as the Queen Mary salute, which required all types of flips, tosses, twists, and turns of the weapon before resting it back on our shoulders. It looked great so long as all of us did it in unison. I remember our first attempt at it. First Sergeant flipped his rifle around this way, then that way, then up and over, smacked it on the ground, twirled it around again, and placed it back on his shoulder. It looked sharper than hell. Then he asked us to duplicate it. He blew on his whistle four times to give us the cadence. Rifles were flying in all directions.

Whenever we were in formation, Averitt, Duthie, Silverman, and I always tried our damnedest to make each other laugh to see if we could get each other in trouble. Duthie and Silverman laughed at anything. They were easy. But Averitt, he could keep a straight face no matter what. The only thing that made him laugh was seeing somebody else get caught. Sergeant always lined us up the same way when we would march in ranks. Averitt was right next to me, and Duthie was behind me. Silverman was beside Duthie. Sometimes, as we marched along, Averitt and I would do a double step or dip down just enough for Duthie and Silverman to notice. They would start laughing. Duthie would usually poke me in the back, and Silverman would say something like, "Damn, you two assholes are gonna get us in trouble."

By that time First Sergeant Thomas would have noticed the abnormality and been right up beside us. "Shut up, Silverman!" he would yell. That would make us laugh even harder.

There was this one little guy—his name was Baglow, but we called him Squirt because he was so short. He looked like a miniature version of Lou Costello, and he talked with sort of a high-pitched voice, like a kid. Anyway, Squirt was always positioned directly in front of me in the ranks when we marched. Because he was such a squirt, his rifle, which he had resting over his shoulder, pointed right up into my face. Whenever we'd do an about-face, he would hit me in the head with it. Finally one day I just smacked the damn thing out of his hands.

The rest of us kept marching along in precise formation as Squirt ran off to retrieve his weapon. First Sergeant Thomas came running up. "What the hell's going on here?" Squirt mumbled something as he scampered back to the formation, and then he started skipping back and forth until he landed on the right foot of the cadence. First

Sergeant got in stride right next to me. "Sacco, what's the problem here?"

"Squirt keeps hitting me in the damn head with his rifle, First Sergeant."

Trying to conceal the fact that he had a slight smile on his face, he moved up the ranks next to Squirt. "Baglow?" Squirt looked up at him with eyes bugging out. "Baglow, you hit Sacco in the head?"

"Uh . . . I don't know."

"Hold your rifle at a steeper angle, Private. That soldier behind you is U.S. Army property. If you injure him, I'll have you arrested for damaging Army property. Clear?"

"Yes, First Sergeant. Affirmative. Roger that, First Sergeant," Squirt answered in the most military-sounding voice he could muster.

"Lots of brass up there in those reviewing stands, Baglow!"

Squirt looked toward the reviewing stands. That's the point where First Sergeant lost his temper. "What the hell are you looking at? You keep your eyes straight ahead at all times! What the hell's wrong with you, Private? You're not at a damn picnic!"

"Yes, First Sergeant."

"We want to see perfection marching in these ranks, soldier!"

"Yes, First Sergeant."

"We want to see precision and professionalism out here, not some little two-bit buck private running around doing a comedy routine looking for his weapon!"

"Yes, First Sergeant."

"You trying to make me look stupid in front of the brass, Baglow?"

"No, First Sergeant."

"Because if you are, I'll ship your ass back to basic, and you can spend the whole war shoveling shit for all I care, 'cause I don't give a damn!"

"Uh . . . yes, First Sergeant . . . uh . . . no, First Sergeant." I couldn't see them this time, but I suppose Squirt's eyes were bugging out again, because he didn't know which way to answer.

The rest of us were trying as hard as we could not to laugh. The sergeant turned, pointed to me, raised one eyebrow, and said, "Watch your head, Sacco," before returning to the back of the column.

Squirt loved the idea of being a soldier. The story was that he had lied about his age because he was too young to get into the military.

Why somebody would do that, I don't know. I also didn't know how old he was, and I didn't care. I just thought he was too damn short to be in the Army.

One day, while we were on one of those twenty-mile hikes through the woods carrying full gear, Squirt turned around to me and said, "Hey, Sacco, get my pack off my back."

"I'm not getting shit off your back, Squirt. What the hell's wrong with you?"

"What's wrong, Squirt?" Averitt asked. "You want one of us to pick you up and carry you the rest of the way?"

"Shut up, Averitt! Come on, Sacco," Squirt said. "I can't carry this damn thing."

"Hey, Duthie," I said, "this little Squirt up here wants you to carry his backpack."

"Tell him to go to hell."

"Silverman, you want to carry Squirt's pack?" I asked.

"Nope."

"Sorry, Squirt. I can't find nobody back here willing to carry your shit."

"Oh, come on, Sacco," he pleaded. "You're big. You carry it, just for a while."

"Why the hell can't you carry it yourself?"

"Because this thing is too big and . . . and I'm too damn little!"

All of us burst out laughing. Averitt said, "Squirt, why the hell didn't you just stay home in the first place and play army in the backyard like all your other little friends?"

Squirt was upset. "All you guys can go to hell. I'll carry my own damn pack." We had a good time laughing at that squirt.

Before I knew it, November was upon us. For some reason I'd always pictured Texas as a warm place, sort of like it was in the cowboy movies. But this part of the state, up in the northeast corner near Oklahoma, was freezing. The northern winds, blasting unimpeded across the vast Texas plains, swept into Camp Maxey without apology. The trees were leafless, the grass was brown, the air was frigid, and all the birds had vacated the area for destinations south. The ever-shortening days were gray, and the relentless wind seemed to have teeth that bit into any exposed skin.

After work was done, most of us opted to sit around inside the relative warmth of the barracks reading or writing letters, talking, or just goofing off. I still read and wrote Chandler's mail for him, so I got to know what was going on in his life pretty well. Still, I could never figure out what he had done to get into prison, and I sure as hell didn't want to ask him. I did try to help him learn to read a few words, though. And I think it helped. I decided to teach him the most important things first, so I pointed out the word "Mom" on his letters. I would say, "Mom. That's your mama. See? M-o-m. Mom." And then I'd show him the word "Becky" and tell him what that meant. He caught on pretty quickly.

Silverman noticed me writing a letter for Chandler one night, and he came over to me. "Sacco, why are you always reading Chandler's mail and then talking to him about it?"

I didn't want to embarrass Chandler, so I said, "Well, he has this girl back home, and he wanted me to help him write her some nice love letters." I had just made that up off the top of my head, so I didn't know if I'd been too convincing.

Silverman looked at me with that curiously bespectacled Silverman look for a second and then said, "Hmmm. That's a good idea. I want you to help me write a letter to my girl, too." He took off for his bunk and brought back a stack of letters from home. "Her name is Rosanna," he said. "You'd like her. She's Italian." He pulled out a picture. "Here. Take a look."

"Oh, wow," I said as I looked at the photograph. "This is not your girlfriend."

"Whataya mean?" he asked. "Sure it's my girl!"

"Silverman, this girl's too pretty to be seen with you."

Silverman knew that was a compliment, so he just laughed. "Yeah, you're right," he said.

Over time I got to know everything there was to know about Silverman and Rosanna. She knew about me, too, because he took a picture of the two of us and sent it to her in New York. She would send him cookies and sausages and other Italian food. Fortunately, she'd almost always include a little in there for me as well.

I was sitting on my bunk writing a letter one night just before lights-out when Silverman came over and sat next to me. He positioned himself so that none of the other boys could see that he had a bundle of cloth in his hand. "Look," he whispered. "See this?"

Since he was whispering to me, I whispered back. "Uh-huh."

"Look at this." He leaned toward me and folded back the cloth, revealing a small, flat box, which he slowly opened. I was dazzled as jewels sparkled from within.

I glanced around to make sure none of the other guys were taking notice of us as I moved closer to Silverman. "What the hell is this?" It was hard to whisper.

"Okay," Silverman began. "I'm a jeweler, you know that."

"Yeah."

"These are some pieces that I have saved for Rosanna."

I looked in the box and reached to touch the jewels. "Is it okay to touch 'em?"

"Yeah, it's okay," he said. "That's what jewelry is for."

There was a diamond necklace, a gold bracelet, and a gold chain with a St. Christopher medal on it. And there were three beautiful rings. "I made the rings," he said. "Designed them myself." Silverman quickly checked the room again to make sure we weren't being watched. Then he continued. "These are all for Rosanna, and I want to make sure that she gets them."

"Well, you need to send these things by registered mail," I instructed.

"No, that's not what I mean," he answered.

"What *do* you mean?"

Silverman studied my face for a moment and then spoke. "I want you to keep them with you until after we get home."

"Huh?" I was instinctively shaking my head.

"Come on, Sacco."

"No way in hell, buddy," I said. "How am I gonna keep up with that stuff? What if I lose it? Nope."

"Look, Sacco, you're the only one I trust." He swiftly scanned the room again. "You take these and, hell, if you lose them, then you lose them. But if I keep them, sooner or later one of these assholes is gonna find out I got them and steal them from me. I think Chicago already has an idea I got something. Nobody's gonna steal anything from you. You just take these and—"

"I don't know, Silverman. It makes me feel a little uncomfortable. Why don't you just send them to Rosanna now? That way she can be thinking about you when she wears them." I thought my logic made pretty good sense.

"No. Look," he said, picking up one of the rings. "Look at this. This is an engagement ring. I want to give it to her in person. You can't send an engagement ring to your girl in the mail! And this one, this is the wedding ring. I kinda have to be with her to give her this."

I pointed to the third ring. "What's that one?"

"This one right here, this one is my wedding ring. This is mine."

"Then why don't you just wear that one?"

"Because I'm not married yet, Sacco. That's the point!"

"Okay, okay. Relax. So why don't you just send these to your parents to hold until you get home?" I asked.

"Because I want to have them with me to remind me of Rosanna and to give me courage. Look, Sacco, me and you, we're both gonna make it."

I knew what he was trying to say, but I still felt a little funny about it. After all, these things looked pretty expensive, and I was sure I couldn't replace them if I lost them. Besides that, I had enough crap to keep up with. "I don't know, Silverman."

"I'll tell you what," he said, ready to cut a deal. "When the war is over, you come to New York and I'll make you a diamond ring like you've never seen. I'll design it especially for you. The Joe Sacco ring. How does that sound?"

"I don't know. . . ."

"A big ring, right there, for that finger. With a diamond. Two diamonds. Whatever you like."

"Not too girly-looking," I said.

"Oh, no. Real nice. Classy."

"Well . . . how long do you want me to hold on to these?"

"Until the war is over," he replied. "Until the war is over."

Damn, I thought. *Now I've gotta be responsible for lugging this shit around for the whole war.* Well, at least it wasn't as heavy as Squirt's backpack. In fact, the whole package didn't weigh much more than Mr. Tony Sciatta's nuts, which were still in my pocket.

Cardini was approaching my bunk. Silverman closed the box and rewrapped it in the cloth. "Okay," I said as I took the package and tucked it safely away under my pillow.

"So, you see, that's why you have to put an amplifier on the line if you take it more than five miles," Silverman said aloud, continuing a conversation that had never taken place.

"Hey, drop the shop talk, Silverman," Cardini instructed. "Enough of that during the day. Sacco, I want you to listen to this." He waved a 78-rpm record album in the air. "Got it from the captain." He pulled a small record player from his footlocker, wound it up, and put on the disk. Hiss crackled through the speaker as he lowered the needle onto the vinyl.

"What's this?" asked Silverman.

"'Nessun Dorma,' *Turandot*, Giacomo Puccini; 'Vesti la giubba,' *I Pagliacci*, Ruggiero Leoncavallo. *Great Moments in Opera*."

"See ya, Sacco." Silverman got up and walked away.

"Listen to this, man," Cardini said to me. "This is great stuff!"

The crackling gave way to tinny music and an operatic tenor voice. I don't know who was singing, but he sounded fat. I do know that my uncles would have enjoyed singing along. "Hey, Cardini, why you always make me listen to this stuff?"

"Because you're Italian. I'm trying to give you some Italian culture. You're supposed to know this music. I'm gonna teach you to appreciate this."

"Oh."

The guy on the record kept singing higher and higher notes until I thought he was going to bust a ball. But he never did, because he'd finish one song and then come right back in with the next. Fortunately, nine o'clock arrived, and with it came lights (and record players) out.

As one might imagine, a great deal of our time was spent out in the field stringing phone lines. We were split into groups of six, and each group was given a truck and a driver, who was also part of the 92nd. My group consisted of Averitt, Duthie, Silverman, Chandler, Spotted Bear, and myself, with Tex as our driver. Tex was specially trained on driving and repairing the truck, so that not only could he get us in and out of rough terrain, but he could also fix anything that broke along the way.

The truck had enough room for maybe two men in the cab with Tex. The rest rode in the back, along with our tools, equipment, and cable. There was a platform on top that was supposed to raise us high enough to string the lines in trees, but it really didn't go too high and was mainly designed as a boost up.

92nd Signal Battalion truck, Camp Maxey, Texas, November 1943.

Each of us had to become expert at climbing trees and telephone poles. We had safety belts that we would hook around a pole, then we'd get a grip with our legs, slide the belt up a bit more, move our legs up to get another grip, and keep shimmying like that all the way to the top. After a while it got not only easy but fun. We would have races to see who could get up to the top the fastest. I was like a monkey up that thing. Of course, once we were at the top, we had to drop a line to get the cable sent up. Then we had to use our tools—preferably without dropping them on the other guys—to hook up the

After the war I was an instructor for Signal Battalion training exercises. I'm on the ground in the back.

**Postwar training exercises.
I'm at the far right.**

line or do whatever work needed to be done. The colder it got, the more difficult it was to hold on to the tools. They had given us gloves, but it was tough to clasp a pair of pliers while wearing them.

We would work in groups of two, so that between the six of us, we could set up lines on three adjacent poles at once. One guy climbed, and the other stayed on the ground to feed cable and whatever else was needed up top. I was usually paired with Duthie.

We heard that a guy from C Company named Whatley had fallen from the top of a thirty-foot pole one day and busted his ass. They sent him to the hospital for a week, and the story was that they were going to send him home to rehab for a few weeks.

A couple of days later, I decided I would "slip" and fall. I figured I could land so that I hurt myself just enough to get sent home for a while. It was drizzly and cold, and the pole was slippery. Duthie had sent me up everything I needed and was down by the truck taking a smoke and talking to Tex. I tried to misplace my foot, but I didn't slip. I jiggled my belt, but I stayed in place. I took one foot away from the pole altogether and just hung there, suspended in air. I unlocked my belt. Nothing. No matter what I did, I was still on top of that pole.

Finally I just said, "Hell with it," secured my position, and went on about my business. And damn if I didn't slip and fall all the way down the pole. I hit the ground hard. Duthie and Tex looked over when they heard the thud.

Once he saw that I was alive, Duthie was trying not to laugh. "Sacco, you okay?"

I stood up, wiped the mud off, checked all my moving parts, and said, "Yeah, I'm fine."

004

Tex didn't say anything. He just chuckled and kept smoking his cigarette. I stood there bending this way and that, trying to figure out if I had injured anything. I was somewhat sore, but nothing seemed to be broken. I heard Averitt's voice coming from the top of the next pole. "Did he bust his ass?"

"He's okay," Duthie shouted back.

Appropriately embarrassed, I hitched my belt again and shimmied back up to the top. First Sergeant Thomas came around later to see if I was all right. I told him I was fine except that my feet hurt. I suppose that was because I had landed directly on them to break my fall. Sergeant said that if they still hurt in the morning, to go over and see the doctor.

Later that night, back at the barracks, Averitt came over and sat on my bunk, "Damn, Sacco, you're the only one I know who could fall off a damn pole and not get hurt."

I was taking off my boots and socks. "I know. I landed right on my feet like a cat."

"If that had been anybody else, they'd've busted their ass wide open," Averitt said.

"I don't know. My ass hit, but not too hard, because my feet broke my fall. That's what hurts. They don't feel so bad now, but Sergeant said that if they hurt tomorrow, then I need to go see the doctor and get 'em checked out."

"What's wrong with your feet? You break a bone in there or something?"

"Nah," I said as I rubbed my arch. "They feel okay now. Just a little sore from landing on them so hard, that's all."

"Damn, Sacco. Put your foot straight out."

I don't know what he was trying to see, and I don't know why I accommodated him, but I flexed my foot out on the bunk. "Damn, Sacco," he repeated. "Hey, Cardini, come look at this." He looked back at me. "Put the other one up here. Do they hurt you all the time?"

"No. They don't even hurt real bad right now. Just kinda sore, that's all."

Cardini looked up from his letter, stood, and strolled over to my bunk. "What?"

"Look at his feet," Averitt instructed.

"What about 'em?"

"Look at these damn things! They're flatter than my ass!" All of the guys, and especially Averitt, used their asses as some sort of measuring device. Things were said to be "flatter than my ass," "bigger than my ass," "colder than my ass," or, in general, any known measurable characteristic as compared to someone's ass. In the case of another person, he might be "uglier than my ass," "dumber than my ass," "shorter than my ass," "slower than my ass," or any possible combination thereof. The ass came in handy when looking for a way to gauge something.

So now my feet were flatter than Averitt's ass. Cardini turned and walked away. "Hey, thanks for making me look at Sacco's foot," he said as he returned to his letter.

I was putting my socks back on. It was cold.

Averitt still wanted to talk about this. "You definitely need to go see that doctor. I'll go with you. When you going? Right after inspection?"

"I don't know. I don't even know *if* I'm going. I've jumped off stuff before. It just hurts for a little while. They'll probably be fine tomorrow."

"I knew this guy back in basic," Averitt began. "He was from Oklahoma or Kansas or someplace. Okay. He had flat feet like yours. Flatter . . . than . . . my . . . ass. Okay. The doctor examined him, and they sent *his* ass home. Gave him a 4-F."

"For having flat feet?"

"Hell yeah. That qualifies you for a medical discharge."

That's all I needed to hear. The next day was Saturday, so after morning inspection, I was on my way to visit the doctor. Averitt insisted on coming with me, which seemed sort of strange to me, but he got permission from First Sergeant Thomas, so we went together. All the way there, he kept telling me, "Let me do the talking."

When we arrived, Averitt started explaining his medical opinion to the doctor concerning my condition. The doctor listened patiently as he examined me for injuries. Averitt kept prattling along until he got to what he considered to be his coup de grâce.

"And he has flat feet. Flat as my ass."

The doctor peered over his bifocal glasses at Averitt, then looked up at me. "Let me see your feet again," he said. I propped them up on the examining table. "Hmmm. Okay. I see." He looked around at Averitt. "I'll just take your word for how flat your ass is, if you don't mind." I almost busted out laughing. Then he turned his attention back to me and poked the bottom of my foot with his thumb. "You don't have much in the way of an arch here, do you?"

"No, sir."

"Have any trouble walking, hiking, or anything?"

"No, sir."

"Yeah. It's usually not a problem." He picked up the chart, made a couple of quick notes, and said, "Well, you check out fine. Looks like you were pretty lucky, soldier. You be careful up on those poles from now on."

"Yes, sir." There was an awkward silence for a couple of seconds, during which everyone was motionless. "But what about my feet?"

"What about 'em?"

"I thought you said they were flat."

"Well," he began to explain, "they are, but it's—"

Averitt was unable to contain himself. "He can get a 4-F for that."

The doctor looked around at Averitt. "He can?"

"Oh, *hell* yeah," Averitt informed him.

"Well, in his case, I think there won't be a prob—"

"I knew this guy in basic, from somewhere—Nebraska or someplace—who had flat feet, and they found out and made him 4-F. Sent his ass home. Sir."

The doctor was bobbing his head very slowly. He pursed his lips and made some notes on a clipboard. "I certainly think this could warrant further examination," he said. Averitt's eyes lit up. "Okay, men," the doctor instructed, "you stay here. I'll be right back." He left the room and shut the door behind him.

Averitt silently gave me the thumbs-up. Within a few seconds, the door opened and the doctor walked back in. "Okay. You can go right over to Dr. Williams's office. That's directly down and across the hall from here." I jumped off the table and buttoned my shirt. "He's expecting you straightaway."

As we were walking down the hall, Averitt said, "You must be luckier than my ass." I didn't know what he meant by that and didn't want to know.

When we got to Dr. Williams's office, he took us right in. "Please, sit down," he said, pointing at two chairs facing his desk. "Let's see." He was looking at some notes on his desk. "Private Averitt and Private

Sacco, is it?" We each raised our hand when he called our names. "Good," he said. "Did I pronounce your names correctly?"

"Yes, sir."

"Good. Then let's continue. I thought we might be able to talk about your situation."

Averitt began, "He can get a 4-F because—"

"So, Private Sacco," the doctor interrupted, "is it okay if I ask you a few questions?"

I nodded.

Averitt, undaunted, continued to talk. "There was this guy from . . . from . . ." He turned to me. "Where the hell was that guy from?"

"I think he was from Arkansas," I said, trying to help.

"No, I think he was from someplace like Kansas or Nebraska or—"

"Look, gentlemen," Dr. Williams said firmly, "as interesting as this is, let's please move on. Now . . ." He cleared his throat and put on his glasses before continuing. "So, soldier, how do you feel about the Army?"

I was perplexed, but I tried to answer as best I could. "Uh, it's okay."

"You made some buddies here so far?"

"Yes, sir."

"You have fun being with your buddies here?"

"Not really fun. They're okay. Sometimes we make each other laugh in formation."

Averitt kicked my foot and shook his head disapprovingly at my answer. The doctor gave a slight chuckle. "That's okay," he said. "That's normal."

I looked over at Averitt and made a face.

"You miss home?" he continued.

"Yes, sir. Sometimes."

"You have a girl back home?"

"Not one steady one."

He wrote something down on his pad, then looked back up at me. "So you don't have a girlfriend back home?"

"No, I don't guess so."

"What about you, Averitt?"

"Yes, sir. Oh, hell yes. I got lots of girlfriends."

Dr. Williams turned his attention back to me. "You like girls?"

I thought that was a funny question. I couldn't figure what it had to do with my feet, but I answered anyway. "Yes, sir."

"What about boys?"

Now I was really stumped. "What about 'em?"

"You like them?"

"I guess so."

The doctor perked up, and even Averitt jerked his head around and looked at me funny. "You do?" the doctor asked.

"I guess so," I explained. "Unless he's an asshole or did something to me."

"Well," Dr. Williams continued, looking for some clarification. "What about your buddy Averitt here?"

"What about him?"

"You like him?"

"He's okay. Talks a lot."

"And that bothers you?"

"When I'm trying to sleep, yes, sir."

"You ever want to touch him?"

This was one question I couldn't understand. "Uh . . . no."

"Well, what about some of the other boys in your barracks?"

"What about them? They all seem to be pretty good guys. They're okay."

"You ever think about any of them?"

"I think about whoever's supposed to be sending me cable when I'm up on top of a pole."

"Interesting."

Averitt was getting anxious. "You see, Doctor, he slid off a pole. Slid all the way down. How the hell he didn't bust his ass, I don't know."

The doctor continued as if Averitt weren't even in the room. "What I mean is this: Do you ever think about some of the other men . . . when you see them in the showers?"

"Huh?"

Averitt started laughing and said, "Oh, hell."

I said, "What the hell are you talking about? I ain't no queer or nothing like that!"

"Calm down, soldier," said the doctor in his most soothing voice. "Calm down. It's okay. We just have to ask these questions at the beginning."

I was wondering, *The beginning of what?* and *What does this have to do with my damn feet?*

Averitt leaned over to me and tried to whisper without moving his mouth. "It's all part of the discharge." It didn't even come out as a whisper, so I'm sure the doctor heard every garbled word. There was a brief silence before Averitt said out loud, "Doctor, this guy has feet flatter than my ass." He kind of trailed off at the end because the doctor looked at him and raised an eyebrow.

Doing his best to ignore Averitt, Dr. Williams turned his attention back to me. "So, Sacco, you think you would like to go to bed with your sister?"

I didn't know what he was getting at, but I had heard enough. The doctor had been agitating me little by little as we went along, and now I was furious. "Hell no," I yelled as I jumped from my seat. "I don't even have a sister, but if I did . . . hell no! What the hell's wrong with you?" I grabbed the back of my chair and picked it up.

Averitt leaped to his feet and grabbed the other end of the chair. "What the hell are you doing, Sacco? Pipe down! Just answer the damn questions. Hell! You're gonna blow your 4-F! And put this damn chair down!"

Dr. Williams had retreated to a corner of his office. "Just put the chair down, son," he said, once again employing his most calming voice.

Averitt said, "It's okay, Sacco. He don't think you're a queer or a preevert or anything like that, do you, sir?"

"No, no, of course not," the doctor replied as he moved back toward his desk. "I just have to ask these standard questions, that's all."

"Put it down, Sacco," I heard Averitt say.

I put the chair back on the floor and sat in it. Averitt and the doctor sat down, too. After a couple of seconds, Averitt said, "Sacco here, his feet are—"

"I know," said the doctor. "They're flatter than your ass."

"Well," Averitt began with eyebrows raised, "I knew this guy in basic—"

"Okay, you two go back to your platoon."

Averitt didn't want to give up. "But what about his feet?"

"He should take his feet with him. Dismissed." And with that, Dr. Williams jumped from his desk, came around to where we were, and escorted us out of his office.

As we were walking back over to the barracks, I asked Averitt, "What the hell kinda crap was that? That damn doctor didn't even ask me about my feet."

"What the hell was what? Damnit, Sacco! What's wrong with you? You must be crazier than my ass! You picked up a damn chair in there!"

"Yeah, and I should've cracked it over his head, asking me if I wanted to go to bed with my sister. I don't even have a sister!"

"Well, he didn't know that."

"I should've cracked both of y'all over the head."

"Damn, Sacco. You been hanging out with Chandler too much."

The winter of 1943 was, to paraphrase Averitt, shaping up to be colder than my ass. Workdays were as long as ever, but sunlight was in short supply. We left the barracks in the dark each morning and returned in the dark each afternoon. Morning roll call in the blasting wind could be brutal, especially after you'd emerged from a relatively warm bed. Each morning when I heard reveille on the loudspeaker, I had to decide whether it would be easier to brave the weather or buck the system.

It was absolutely freezing one early December morning, even colder than normal. I hadn't been able to sleep too much the night before, probably because I was homesick and worried about the war. So when reveille sounded, I said to myself, "Hell no. I'm not getting up."

All the rest of the guys got dressed and went outside for roll call. After a few minutes, First Sergeant Thomas came strolling into the barracks. "Sacco? You getting up?"

I was completely under the covers and didn't want to stick my head out. "Nope."

"You sick or something?"

"Nope."

"You want to sleep a little longer while the rest of us get to work?"

"Yep."

"Okay. Well, when you get up, go on over and report to Captain Clark's office."

"Okay."

I got to the captain's office about 0900. I was still sleepy and cold, but now I was also cranky. For some reason I felt like they had no right to make me get up so early on such a frigid morning. In fact, they had no right to make me join the damn Army in the first place. I had been home minding my own business, and before I knew it, I was on some godforsaken Army base doing all type of crap in the cold. As I swung the office door open, I saw a familiar face. "Silverman, what are you doing here?" He was sitting at a small desk stacked with documents.

"I do paperwork for Captain Clark every day, but not Monday and Tuesday. I can type. What the hell are you doing?"

"Aw . . . I don't know," I answered.

Captain John Clark, seated behind a larger desk in the same room, cleared his throat. "Uh, soldier, can I help you?"

I was supposed to salute, but I was too pissed off, so I just said, "Yeah. Sarge said you wanted to see me."

"And you are?"

"Sacco," I said, holding up one of my dog tags as if he could read it from where he was sitting.

"Ah, yes, Private Sacco," he said, ignoring the lack of a salute. "I do want to see you."

"Well, here I am."

Silverman, surprised that I would talk to a captain as if he were one of the boys, hid his face in his hands. Captain Clark continued. "Private, First Sergeant Thomas informed me that you didn't want to get up and join your men for work this morning."

"That's right."

"Why not?"

"Well, first off, I felt like shit, and I was sleepy. Plus, it's colder than a witch's titty out there, so I covered up and went back to sleep."

Silverman turned his head away to keep from laughing.

Clark didn't find it so amusing. "So you just decided you could sleep later than the other men, is that right?"

"That's right."

"Sacco, don't you know that there are boys overseas right now in foxholes freezing their asses off? Do you realize that, soldier?"

"Yeah, that's fine with me."

Silverman was shuffling papers, dropping things, opening file drawers, doing anything to keep from laughing.

"What do you mean, that's fine with you?" Captain asked.

"Okay, when I go overseas, I'll get in a foxhole and I'll freeze my ass off, too. But this morning, I figured what the hell, and so I decided to keep my ass warm while I could."

Silverman burst out laughing, got up, and ran out of the room.

Clark called to him. "Silverman! Get back in here and shut the hell up!" He then turned his attention back to me. "Okay, Sacco, go on over to the mess hall right now and enjoy some KP duty for a week."

I buttoned up my coat, went over to the mess hall, and found the mess sergeant. I told him Captain Clark had ordered me to report for one week's duty. He looked at my face for a couple of seconds and asked, "Are you from Birmingham?"

"Yeah."

"Your daddy is Mr. Jake?"

"Yeah." I wondered how he knew that.

"Hell, I'm Frankie Ranelli! From Birmingham!"

"Oh, hell yeah!" I thought he looked somewhat familiar, but now I realized that he was from one of the Italian families back home and that I'd seen him around town a few times. "Hell yeah! Joe Sacco."

We got along fine. He had me peeling potatoes, washing pots and pans, doing easy stuff like that. Being in the kitchen all the time, I got to eat all the time, which was great. What I enjoyed the most was drinking milk. Frankie told me that I drank too much milk, that it would make me sick, but it never did. KP was supposed to be a punishment, but I actually liked it. The only problem was that it did require me to get up even earlier than normal to help prepare breakfast for the other boys. I didn't like that part. Even so, the only way for me to continue to work there beyond a week would have been to get in trouble again. And I didn't want to do that, because the next time they might have had me cleaning the latrine instead of pulling KP. So after that I got up on time, no matter how miserable I felt or how cold it was.

Of course, some of the guys figured they had the system beat. A few of them had developed a technique based on the fact that our mattresses weren't firm enough and tended to bow down in the middle. Here's the way it worked: One soldier would get right in the middle of his mattress, which would sag down. His buddy would pull the top sheet and blanket tight over the sag and tuck it under the sides of the mattress. If it was done correctly, the bed looked perfectly flat and there was no evidence of a sleeping soldier within. The buddy would then answer for the sleeper at roll call. The next day the two would swap.

Cardini wanted to master this in the worst way. His father was an admiral in the Navy, and Cardini took that to mean that the rules didn't necessarily apply to him—an attitude that more than likely accounted for the perpetual smirk on his face. Unfortunately, he teamed up with a guy named Wilkins who wasn't, let us say, the smartest cookie in the jar.

Wilkins was a scrawny fella with buckteeth and the thickest horn-rimmed glasses I'd ever seen. We used to tell him it looked like he had the bottoms of Coca-Cola bottles up to his eyes. The extent of his intellectual prowess was made clear one night while we were sitting around the barracks. Wilkins was reading the local Texas newspaper. At one point he looked up from the paper and said, "Hey, Spotted Bear?"

"What?"

"I'm reading this story here in the paper."

"Yeah?"

"And . . . what does this mean right here?"

"What?"

"Well, it's talking about this girl who was kidnapped two weeks ago, and it says, 'Officials haven't found her yet.' "

"What's your question?" Spotted Bear asked.

"What does that mean?"

"What does *what* mean?"

"What does it mean, 'they haven't found her yet'?"

"What the hell do you think it means?" Averitt asked.

Duthie tried to be helpful. "It means they're still looking for her. They haven't found her yet."

Wilkins, apparently unsure about the names of all parts of the female anatomy, was still mystified. "But what's her 'yet'?"

The whole barracks, with the exception of Wilkins, erupted into laughter. If that had been the only stupid thing he ever did, I'm sure we would have kidded him about it for the rest of the war. But, true to predictions and form, he would far surpass this in time to come.

Anyway, he and Cardini ran their sleeping scam a few times, but then Wilkins started messing things up. He would never correctly make up Cardini's bunk, so that when the corporal inspected the barracks after they were supposedly empty, he would invariably notice the abnormality, and Cardini would get nabbed. Then, even if the bunk was made up perfectly, the squeaky-voiced Wilkins could never accurately impersonate the deep-voiced Cardini at roll call. So Cardini was never going to get away with it, which pissed him off no end, especially since he considered himself the king of getting away with stuff.

As for Wilkins, even though Cardini always made his bunk and covered for him at roll call, he seemed to find a way to screw things up. The most famous time was when the corporal was inspecting the barracks and everything seemed to be in order, but then, as he was about to leave, he heard a loud, extra-large fart. He walked back down the row of bunks to investigate the origin of the sound. Just as he neared Wilkins's bunk, he heard another one, this one even louder. He pulled back the covers to find Wilkins trying his best not to laugh but suffocating in his own stink. The corporal grabbed the edge of the mattress and flipped the bunk over, rolling the farter—and his gaseous emissions—onto the floor.

As distasteful as it was to awake in the predawn darkness and then go stand outside in the cold, it was by no means our most pressing concern. There was beginning to be lots of talk among the boys as to when and where we would be shipped out. We all knew that the time was near, but nobody knew for sure if we would be going to the Pacific or to Europe. Everybody seemed to have some insider information, but the truth was that none of us really had any idea.

Averitt had said that we would be getting our steel helmets soon. The Army issued steel helmets just before shipping soldiers overseas. There was one company of guys who were sent to a place called Jefferson Barracks, where they were issued their helmets and then shipped out to the Pacific. Word had it that we would be going to Jefferson Barracks, too. At least that's what Silverman said. I guess he overheard things in the captain's office.

None of us liked the idea of going to the Pacific. It wasn't that the European theater of operations was any safer than the Pacific, but it just seemed closer and more familiar to us. They say that war is comprised of one surreal moment after the other, millions of them all strung together until nothing is real anymore except for one's own mortality. It was strange and more than a little eerie—especially since we all talked about how we would make it home someday—but deep down we felt that if we had to die, we didn't want to die so far away in the Pacific, that, given the choice, we would rather have our lives end in Europe.

Then again, it's not like we had a choice. And, for that very reason, it would have been better not to even think about it. But it was difficult not to think about it when everybody around me was talking about it. The whole issue was making me feel quite helpless, like I had no control over what I would do, where I would go, and where I would die. It was fatalistic and morbid, but a lot of the guys felt that way. Being a soldier during a world war doesn't exactly foster feelings of warmth and security.

Adding to the consternation around the camp was the fact that Christmas was about two weeks away. If you wanted to bring on a major case of the red ass, all you had to do was mention the word "home" or "Christmas" to any of the guys. To make matters worse, Cardini pulled out his record player every night and played Christmas music, which made us feel good and bad all at the same time. Whenever he put on "I'll Be Home for Christmas" or "White Christmas," you could see some of the boys in the barracks cover their eyes. It was tough. This was the red ass in the worst way.

I got assigned to guard duty one night with Hodges. We were to pull the 2:00-to-4:00-A.M. shift, and, as one might expect, neither of us was too happy about it. At the appointed time, Hodges and I got up, dressed as warmly as we could, and went out to our post. It was drizzling, cold, and a bit foggy. As the night progressed, it got colder and colder, until the drizzle turned to sleet, making thousands of tiny sounds as it struck the ground. Ice was forming along the telephone lines and on the bare branches of trees, and they were beginning to sag lower and lower. Hodges was humming one of those Christmas songs real quietly to himself.

I lit up a cigarette. "Ain't this shit?"

"Yep."

"Sergeant said that after 0300 we could go over to the mess hall and get some coffee."

Hodges shook his head. "This is a bunch of shit."

"Yep."

We had an assigned area between several barracks where we would patrol. I didn't know what we were looking for, considering we were in the middle of a heavily guarded Army base, but here we were. We had weapons but no ammunition, so I'm not sure what we would have done if we really had needed to fight somebody. I must have smoked about a hundred cigarettes that particular night. There wasn't much else to do.

We weren't supposed to talk, but I didn't like it to be quiet for too long, because I would start thinking about going to the Pacific or to Europe, and I would start wishing I was home in Birmingham. So I looked for something positive to say. "First Sergeant told us that maybe we could take off one hour early tomorrow afternoon. Maybe get some shut-eye."

Hodges didn't respond but just kept humming to himself.

"That would be good," I said in response to my own statement. "I'd like that."

"You know what?" Hodges asked me.

"What?"

"This is bullshit."

"Yeah. I know."

We were walking between some of the barracks near the edge of the camp when we heard a strange sound. It was faint at first, then got louder. Hodges and I stopped in our tracks. The sound got more and more distinct, almost rhythmic, like it was coming right at us. We could see nothing, but we could hear the clank, clank, clank getting louder. We stood motionless, listening carefully, peering into the darkness, doing our best to discern what was happening. Then out of the mist appeared a company of the 104th Infantry Division, emerging like ghosts in the night, marching out into the field with full pack and gear. Not a word was spoken as they silently clanked past us on their way to the bivouac areas, each man looking into our eyes as he went by. I put my cigarette out on the frozen ground.

Hodges and I stood there and watched until they completely disappeared into the mist again. Then we turned and continued our rounds. We both knew that as bad as we thought we had it, the poor infantry always had it worse.

On December 15, 1943, just one day after we had been issued our steel helmets, First Sergeant Thomas woke us with the following words: "Rise and shine, boys! This is what we've been training for. Get your asses out of bed; we move out tonight at 1800. Somebody tell the Führer we're coming to kick his ass!"

Chicago jumped up and yelled. Chicago was fond of yelling. A lot of the boys in our barracks cheered. After all our training, I guess they were anxious to get into the war. They were also happy to hear that we would be fighting in Europe. I was happy, too, but I didn't exactly feel like cheering. I got up and, as instructed by First Sergeant Thomas, began packing my personal items for the trip. Before we left, I pulled out a little notebook I kept and wrote the words "Well, time to go fight a war."

That evening we boarded a train from Camp Maxey, Texas, to Camp Shanks, New York. The trip took three days, during which time there wasn't a whole hell of a lot to do other than look out the window, play cards, or maybe read some old letters from home. Averitt sat next to me the first few hundred miles and talked my head off, so I tricked Duthie into sitting there, and I went back to another car. Silverman was excited about going to Camp Shanks because it was just a few miles from New York City, and he was hoping he could see Rosanna while we were there. Spotted Bear and Sam Martin slept most of the way. Chicago spent most of his time doing what he did best—that is, smarting off to people. Unfortunately for him, he incurred the wrath of Chandler when he called him a "big, dumb-ass jailbird." Chandler grabbed him and lifted him off the ground, one hand to the nuts and one hand to the throat. Chicago was pretty quiet for the rest of the trip.

5

Shipping Out

Camp Shanks was a staging area, a point of embarkation, where we would be processed before shipping out to Europe. Located across the Hudson River just north of New York City, the camp had the austere look of any Army facility, but with an unusual feel of controlled chaos. The first thing I noticed when I got off the train was that it was warm, probably at least 60 degrees. I had always heard that New York was a cold place, but it was certainly a lot warmer here than in Texas. I assumed it must have been a mild southern breeze blowing up from Alabama to warm our final days Stateside. Whatever it was, it felt good for a change.

We were told that we would be here for only a few days, but, because of the season, those days necessarily included Christmas. We had trained hard, and now we didn't have much to do, so First Sergeant Thomas arranged for us to get passes to go into town and sightsee if we wanted. A group of us decided to explore New York City on Christmas Day.

Averitt, Duthie, Silverman, Chandler, Cardini, Hodges, Spotted Bear, Sam Martin, and I were basically a bunch of hicks in the big city. Well, Silverman was from here, so he wasn't really a hick. His girl, Rosanna, met us at Grand Central Station and acted as our tour guide for the day. She was real pretty and Italian-looking, with dark hair, brown eyes, and rosy cheeks. She seemed happy to have nine soldiers following her around. We were, of course, sporting our Class A dress uniforms, so we did look sharp. But most of us must have looked like country bumpkins, pointing up at every tall building and getting Rosanna to take pictures of us posing in front of every statue in town. I had never seen so much concrete in my life. We went to Times Square, Broadway, and Radio City Music Hall, where we took the tour. The most amazing sight to me was the Empire State Building. We could see it from a few miles away, but when we finally got right underneath it and looked straight up, all any of us could say was, "Damn."

We went to a place called Jack Dempsey's Restaurant in Times Square for Christmas dinner. Jack Dempsey had been a great boxing champion back in the late 1920s. Of course, all of us agreed that even at Dempsey's best, Rocky Graziano could've kicked his ass into next Saturday. Anyway, Dempsey owned the place, so I guess we thought that maybe he would show up and wait on us himself. He didn't, but there was a picture of him hanging on the wall behind the cash register.

When we walked in, people clapped and waved to us. That felt good. I guess they knew we were on our way to the war. I had always heard that New Yorkers were unfriendly, but they were real nice to us. The manager came out and pushed a few tables against one another so that our whole group could sit together. There were a lot of people in there, which I thought was strange for Christmas Day. I figured that everybody would be home, but Rosanna said that people go out on Christmas in New York, and Silverman said that a lot of people in New York were Jewish, so they didn't celebrate Christmas. For whatever reason, it was fun to see so many people out having a good time.

Toward the end of the dinner, when the waiters were taking the plates off our table, I realized what a nice time we'd had, and I started feeling sad that it was coming to an end. After all, it was Christmas Day, and the truth of the matter was that I was a long ways from home and about to be sent even farther. A group of us had gone to church early that morning back at camp, and the priest had told us that everybody in the United States would be praying for us while we were away fighting. I had been pretty homesick, but I guess being at Christmas Mass and hearing all the prayers had made me feel a sense of peace. And now sharing a good meal with my buddies helped me forget how scared and alone I was.

We left Camp Shanks in the early-morning hours of December 27, 1943, for the Forty-fourth Street docks in New York City, where, after the usual Army delays, we boarded a ship called the USS *Anne Arundel*. I didn't know who Anne Arundel was, but I had a cousin named Annie Renda, and that sounded about the same, so I took the name of the boat as a good sign. I had never seen a ship in real life and therefore didn't have a point of reference, but this thing definitely looked huge. In fact, the first thing I thought was that it looked much too big and heavy to float.

When I got to the top of the gangplank and set foot on the ship, I was amazed at how steady it was. It felt just like being on the ground. I guess I expected my weight to cause the ship to shift around in the water. Of course, the only boat I'd been on previous to this had been a little two-man canoe at East Lake Park in Birmingham.

Many civilians had come to see us off. Rosanna was there, as was Silverman's entire family. There were a priest and a rabbi, both blessing the ship, so we had all our bases covered. When we got up to the top and looked out over the railing, we could see a buzz of activity on the docks below, with trucks and equipment everywhere and hundreds of soldiers streaming up the gangplanks onto the ship, duffel bags on their shoulders, waving to their loved ones. All the activity began to mesmerize me until it seemed almost peaceful. As I leaned against the railing and watched the scene unfold below me, it crossed my mind that some of these men might not return home alive. I tried not to think about it, but then Averitt said, "You know what? Look at all this shit we have to go through and get ourselves killed all because some little asshole Hitler can't mind his own damn business!"

None of the boys knew anything about our route or even where we were actually going, other than to Europe—probably England. But First Sergeant Thomas had told us that we didn't need to know or even speculate on the trip because, as the saying goes, "Loose lips sink ships," and though it was great to see everyone come out and see us off, it was possible that there could have been least one or two Nazi informants out in the crowd.

The 92nd was one of the first battalions on board that morning, but it took most of the day for everybody to get loaded and for the sailors to ready the ship to leave. We were finally towed away from the docks at about 1700 (5:00 P.M.) amid the waving and tears of the civilians on the docks.

Silverman had given Rosanna the engagement ring in person, as he'd desired. Now he waved her wedding ring in the air as a promise that he would return to her. His parents were crying and holding out their hands to the ship as if to embrace their son. All of us who were together there at the railing waved back to them. I felt like they were taking the place of my parents and that Mama and Papa, if they had been there, would be acting the same way. So all of us—Averitt, Duthie, Cardini, Chandler, Spotted Bear, Sam Martin, Chicago, Tex, myself, all of us—smiled and waved and reached our hands back to the Silvermans as if they were our own families.

When we couldn't see them anymore, Silverman turned to me and handed me the wedding ring. "Here, put this back with the other jewelry."

I took the ring and put it in my pocket. "I wish you'd take all this stuff back and keep it with you," I told him.

"No, Sacco, you hold on to it for now."

I didn't understand something. "Why didn't you give Rosanna that other jewelry? All you gave her was the engagement ring and the St. Christopher medal."

"Those were for her engagement," he answered. "The bracelet and necklace are wedding gifts. You hold on to everything until then. Don't worry, you're invited to the wedding."

"You better hope I don't lose this stuff, Silverman."

He said, "Okay," and just kind of laughed. He had more faith in me than I did.

First Sergeant Thomas told us it was time to eat, so we went belowdecks to the dining area. It had the feel of a big gray cave crowded with hundreds of soldiers and sailors all talking at once. The food was not good, but tolerable enough for me to get it down. Partway through supper we could hear the big engines starting up. The whole ship began to shake and rumble from deep within. For almost two hours, we had been towed out into the harbor. Now, slowly, the *Anne Arundel* was moving under her own power.

We were stowing away our gear after dinner when some of the guys called us to come up on deck. It was dark now and getting cold, but what we saw was beautiful. Lower Manhattan was gleaming in the night, and just in front of it, just to the left, was the Statue of Liberty. When Papa arrived from Sicily in 1921, he had looked upon this very statue as his ship sailed into New York Harbor. He said that all the immigrants crowded the decks because they wanted to see her and remember that moment. Now we were passing by the same place, gazing at the same lady, but we were going off to protect this country. I never thought I would see anything as beautiful as the Statue of Liberty. And I bet that Papa, when he saw her way back then, never imagined that his son would one day go off in the opposite direction to defend her.

It was getting colder as the ship moved out to sea, but despite the frigid temperatures and the strengthening wind, we stayed on deck as long as we could see the twinkling of the ever-decreasing New York skyline on the horizon. For all of us, there was an awesome reality to that moment—the feel of the powerful ship plowing through the darkness of an even more powerful ocean, carrying us into the unknown. The one thing we did know was this: Our destinies awaited us on some distant shore, while America, our home, was quickly disappearing into the cold, wet blackness of night.

Pre-invasion bivouac outside Oxford, England, May 1944.

PART II

Staging for War

When I want my men to remember something important, to really make it stick, I give it to them double dirty. It may not sound nice to some bunch of little old ladies at an afternoon tea party, but it helps my soldiers to remember. You can't run an army without profanity . . . and it has to be eloquent profanity. An army without profanity couldn't fight its way out of a piss-soaked paper bag.

As for the types of comments I make, why sometimes I just, by God, get carried away with my own eloquence.

—GENERAL GEORGE S. PATTON

6

The Voyage Beyond

Aboard *USS Anne Arundel*
The North Atlantic's "Torpedo Alley"
December 28, 1943

It was deep into the night when I awoke to hear a strange creaking and moaning of the ship, as if she were struggling to stay afloat. I wasn't a sailor, so I really didn't know what was going on. What I did know was that the whole boat was rocking this way and that, up and down, back and forth. We were in these little bunks that swung out from the wall and were stacked about four high, with just barely enough room to turn over without hitting the one above you. Even so, I was hoping I wouldn't get flipped out onto the floor. Life's tough enough without the damn room moving all over the place.

I could hear a few of the guys making those seasick noises, but despite the constant motion of the ship, I really didn't feel queasy or anything. In fact, in a strange way all the gyrations were surprisingly fun. It was as if we were on a big ride at the fairgrounds. What concerned me was the idea that we might sink. That was the main reason I hadn't joined the Navy in the first place—sailors spend a great deal of their time on the ocean. And the ocean consists of a lot of really deep water. This happened to be the middle of winter, and we were somewhere in the North Atlantic, which meant that this particular water had to be quite frigid. In addition, there were Germans prowling around in submarines looking to blow us up before we could even get into the war. I thought about all of that for a while. Then I figured what the hell and went back to sleep.

The next morning they blew a whistle to wake us up. Now I had another reason I was glad I hadn't joined the Navy. At least those of us in the Army got to hear a bugle in the morning, not some silly whistle. I felt like I was in one of those Popeye cartoons. Anyway, I slid out of the bunk to find Averitt and Cardini standing there with sallow faces and bloodshot eyes.

"What's wrong with you?" I asked.

Cardini made a groan, his customary smirk nowhere to be seen. Averitt spoke. "How the hell did you sleep?" His eyes were greener than normal.

"Fine."

"No, I mean, how in the hell were you able to sleep through all this?" he asked as he pointed around and simulated the motion of the ship with his hands. The ship, by the way, was still moving, heaving, up, down, backward, up again, sideways, down.

"Oh, I don't know," I answered. "Hey, let's go eat."

"Time to eat?" Spotted Bear asked as he slithered from his bunk. "I'm in."

"How the hell . . . ?" Averitt held his stomach. "Even the thought of food makes me want to puke."

"Good," Chandler said. "More for me. Let's go, Sacco."

"Hey, Cardini," Silverman called out, "I thought your papa was a big-time Navy admiral."

"Yeah, he is," Cardini answered.

"Well, then what's your problem?"

"Shut up, Silverman. Whataya mean, what's my problem? If I could walk without throwing up, I'd come over there and—" Cardini covered his mouth with his hand.

Sam Martin couldn't resist. "Come on, Silverman," he prodded, "this could be your one and only chance to actually kick Cardini's ass."

Silverman made a face. "Ah . . . I feel sorry for the poor seasick bastard. Besides that, I'm hungry. You stay right here and puke your guts out, Cardini, and I'll come back and kick your ass after I eat."

It was easy to find a place at chow that morning. That was the good news. The bad news was that the food tasted like crap. As far as I was concerned, that was yet another reason not to be in the Navy. Spotted Bear wolfed down his breakfast like . . . well, like a bear. Silverman, still working on a pile of something that resembled eggs, asked, "So whataya think?"

Spotted Bear leaned back in his chair and belched out loud. "Tastes like they cooked all this shit in seawater."

"Yeah, seawater that somebody pissed in first," Sam Martin added.

Duthie looked up from his plate and said, "Hmmm . . . if the damn rocking don't make you want to throw up, the chow will." He then lowered his head and continued eating.

I couldn't easily identify everything on my plate, but I did end up eating all they put in front of me. Regardless of the movements of the ship or the taste of the food, I was hungry.

A lot of the guys stayed either in their bunks or at the railing up on deck the first day or so. But most of them eventually came around. I suppose it took some time before their systems got used to all the motion. I was lucky I never got seasick, despite the fact that the ship did a lot of moving, especially in the first few days at sea.

On December 30 we reached a designated rendezvous point about fifty miles east of Boston, where we met up with thirty-five other ships, including the battleship USS *Texas*, three carriers, twenty destroyers, four tankers, and several British troopships. From there the convoy set out on a course somewhere—we really didn't know where. Cardini said we were going to England, but just because Cardini said it didn't make it true.

I had always heard that the ocean was blue, but this one seemed to have no color at all (if you don't count gray). What with the gray water, the gray skies, and the gray ships surrounding us, it seemed like we had sailed smack into a black-and-white photograph. Perhaps I was expecting sunny skies shining over a sparkling clear blue sea as dolphins jumped out of the water, twirled, and splashed alongside us, like I'd seen in pirate movies. Then again, never having seen the ocean in person, I'm not sure what I was expecting.

Despite the lack of color—and dolphins—I did find comfort in the fact that the USS *Texas* was steaming along right beside us. I had thought that the *Anne Arundel* was big, but the *Texas* was one gargantuan boat, and the sight of her so close at hand was comforting. I felt that now I could enjoy the thrill of the waves without having to worry about swimming all the way back to New York in case we flipped over.

There wasn't much to do other than eat, talk, gamble, and think. Of course, thinking wasn't a favorite pastime, because it was difficult to

Where the Birds Never Sing

keep from thinking about the dangers of the coming war, and if we dwelled on that too long, we'd develop a rather large case of the red ass. The food wasn't exactly gourmet, so nobody wanted to eat. That left talking and gambling. And smoking.

Each of us had been issued two cartons of Raleigh cigarettes, which in our estimation tasted like crap. I had never smoked as a kid, but I picked up the habit back at Camp Maxey and now, like all the soldiers, I was a veritable smokestack. But that didn't mean any of us were ready to smoke Raleighs. Fortunately, once we were two hundred miles from New York, we were allowed to buy the better brands—Lucky Strike, Chesterfield, and Camel—aboard ship for five cents per pack. We held on to the Raleighs in case of emergency.

I suppose every other place on earth had New Year's Day, 1944, but we didn't. I'm sure we had the actual day, but it seemed to just come and go without anybody taking notice. I don't even remember anybody saying a word about it. Of course, since we were at sea and were passing through one time zone after the other, it would have been difficult to discern when the clock officially struck midnight. Then again, it's not like we had much to celebrate, considering where we were—and where we were heading. You don't see a lot of parties on a ship filled with men being sent off to war.

The seas became rough during the first three days of 1944, and the ship was once more doing a lot of heaving up and down, which made most of the guys sick again. I didn't mind it all that much, especially as long as I was outside watching the approaching waves and the general motion of the ocean. In the distance, across the choppy waters, I could see the other ships in the convoy. They looked small on the massive, undulating ocean, but they seemed to be steaming ahead in formation with great resolve and purpose.

We started hitting bigger and bigger waves, each lifting the ship like a toy and then sending it crashing down into a valley of water. I was standing at the railing one afternoon when our ship dipped low and got simultaneously hit by a monster wave. It didn't knock me over because I saw it coming and was holding on pretty tightly, but it did get me soaking wet. And, having had the unwelcome opportunity to make a direct comparison, I can honestly say that the water was much, much colder than my ass. In fact, it was freezing.

As the water rushed off the deck and back into the sea, I heard a voice boom up from behind me. "Soldier, what the *hell* are you doing?" It was a Navy officer.

"Just watching the . . ." I heard my voice trailing off.

"Get your ass belowdecks and stay away from that railing in seas like this! You trying to get yourself swept overboard and killed?"

"No, sir," I said. "Just . . . Yes, sir." I scampered away as fast as my waterlogged boots would carry me. I had been having fun, but I did welcome the prospect of going back to my warm, dry bunk and getting changed.

When I arrived, I found most of the guys in various stages of unrest. Silverman was trying in vain to write a letter, but the movements of the ship kept interfering with the movements of his pen. Averitt was reading something and looking several shades of green. Tex was sitting on the edge of his bunk staring straight ahead. Cardini was singing old Italian songs to anybody who would listen. And Chandler was sleeping like a baby.

Spotted Bear entered the area with Duthie and another guy trailing. "Hey, Sacco," he called out. "Hey, what the hell happened to you? You get tossed overboard?"

"Oh, no, I was just up—"

"Hey, look, we don't care," Duthie cut in.

"That's right," Spotted Bear added. "We don't care. Get dried off. It's time to go. This here is Hutter. Hutter, Sacco."

Hutter looked familiar—tall, with olive skin, brown eyes, a Roman nose, and dark hair that came to a point in the front. "Hi, buddy," he said as he shook my wet hand. "I've seen you around."

"Yeah," I answered. "Joe Sacco."

"Joe Hutter."

"Where we going?" I asked as I peeled off my cold, wet shirt.

"Sacco, tell me something," Spotted Bear asked. "You ever shoot dice before?"

"What?"

"Dice. You know what dice are?"

"Yeah, I know what the hell dice are."

"Sure you do," Duthie said.

"Okay, good," Spotted Bear continued. "You ever shoot dice?"

I didn't know what he was getting at. "Uh . . . no. Why?"

They looked at each other and smiled. "Outstanding!" Duthie said.

Hutter chimed in, "He's definitely our man."

"Hurry up and get changed," Spotted Bear instructed. "You're shooting for us."

"What the hell are you talking about?"

Duthie had been digging in my duffel bag, and now he tossed me a dry pair of pants. "Here, put these on."

"Get him some socks outta that thing," Spotted Bear said to Duthie. "Sacco, how in the hell did you get so wet?"

"I thought you didn't care."

"You're right—I don't," he responded. "Come on, we got a craps game to attend, so shake a wet leg."

"But I just told you that I never shot dice before," I protested as I quickly put on a dry T-shirt.

"Yeah, I know," Duthie responded. "We heard you."

Spotted Bear flashed his trademark toothy grin and put his hand on my shoulder. "That's exactly why we want you to do the shooting."

"Whataya mean?"

"Beginner's luck, my friend," he answered. "Beginner's luck."

We won about nine hundred dollars that night. All I did was shoot. Duthie, Spotted Bear, and Hutter made all the bets. I didn't know what the hell I was doing. I'd just throw the dice, and they would holler and collect money. Duthie kept rubbing my head as if I were a genie. Strangely enough, I didn't mind it all that much as long as we were winning. So we ended up with a pile of money, which we split equally among us. But we really didn't have any place to spend it. It's not like the boat had a gift shop.

We went back down there the next night, but I guess I wasn't a beginner anymore, because I couldn't shoot worth a crap. We lost about a hundred bucks and got the hell out. We still felt like winners because we each had about two hundred dollars' worth of extra cash in our pockets. In the Army, that's a fortune.

Understanding how things can mysteriously get "lost" on a ship, I strapped my winnings (along with Silverman's jewels and Mr. Tony Sciatta's nuts) in a neat little package around my waist for safekeeping while I slept. About two nights later, just as I was dozing off, I felt a hand on my hip. I opened my eyes and looked around but didn't immediately see anyone. Within a few seconds, I realized that the hand was coming from below.

I didn't know the guy in the bunk directly beneath me—he wasn't in our company—but he had seemed like a nice enough guy. However, I was sure he knew about the money and probably had seen me strap it on before I went to bed. Now he was after it. It's not as if I could possibly imagine his hand accidentally snaking up along the wall and ending up on me while he was sleeping, but I thought I would lie still and watch him go for the money and then catch him in the act. That way he couldn't make any excuses.

He touched the money belt briefly but then moved his hand away from it. *Hmmm . . . what the hell is he doing?* I wondered. Before I knew it, he had reached down and was trying to unbutton my pants. Now he was going for my *other* package.

I grabbed his arm and jumped straight up, banging my head on the bunk above me before rolling out onto the floor. As I pulled his arm, I twisted it around my bunk and pinned him against the wall. "I'll kill your ass, you damn son of a bitch!" I yelled as I hit the deck.

Duthie, Averitt, Chicago, Tex, and all the guys in the area jumped up as fast as they could. "What the hell's going on?"

"This damn queer was trying to grab my thang!" I blurted out. I was still twisting his arm with all my strength.

Spotted Bear started laughing.

"Your *thang?*" Chicago asked.

"Yeah, my thang," I answered. "My pee-pee, my pole, my rod, my *thang.*"

"He's talking 'bout his dick," Averitt translated.

"That's what I figured," Chicago responded.

"But thanks for the clearing that up, Averitt," Silverman added.

The offender was in pain. "Let me go," he pleaded.

"Like hell!" I said. "Hey, somebody call First Sergeant Thomas!"

Duthie jumped down and looked at the guy. "Sacco, let him go. You're gonna hurt him."

"Hurt him? I'm gonna kill his ass as soon as I let him go!"

Spotted Bear seemed fascinated by the spectacle. "Who the hell is that anyway?"

"I don't know," Averitt said. "But I think his Indian name is Dick Grabber."

Fortunately, the commotion had grabbed the attention of First

Sergeant Thomas, who appeared out of nowhere. "What the *hell* is the problem here?"

"Tell him, Sacco," Spotted Bear said.

I still had the guy's arm twisted around and up over my bunk. "This boy was reaching up here grabbing at my . . . thang!"

First Sergeant looked at me funny, like he either didn't know what a "thang" was or he didn't believe me. Then he rolled his eyes and shook his head, trying not to laugh. "Okay, let go of him," he said. I held on tight. "Let go of him, Sacco." He bent down and peered into the bunk below. "Okay, soldier, what's your name?"

"King, sir."

"Sure it's not Queen?" Averitt queried.

"Looks like a queen to me," Chicago added.

Sergeant Thomas, who was becoming very adept at ignoring our comments, continued to focus on King. "Don't call me 'sir,' soldier."

"Yes, Sergeant."

"What company you in?"

"B."

"That explains it," said Spotted Bear.

"Sacco," Thomas said, "I thought I told you to let go of him!"

I released, and the fellow squirmed away from the wall. The first sergeant pulled the covers from his bunk. "Okay, let's go," he ordered. "Playtime's over. Let's go see your CO."

"Hey, Sarge," Spotted Bear called out.

"What?"

"Don't let him grab your thang."

"Go to sleep," First Sergeant Thomas said as he and the grabber exited the area. It took a couple of minutes for everyone to settle down, since they all had something smart-ass to say to me about what had happened. I didn't care. The sea is a dangerous place, but at least I could now rest a little easier in the knowledge that I wouldn't necessarily be torpedoed from the bunk below.

The seas smoothed out somewhat over the next few days, but I found it increasingly uncomfortable to spend much time up on deck, because the air seemed to be getting colder with each passing hour. Silverman said it was because we were going north, but I couldn't

figure out his logic. "North?" I asked. "I thought Cardini said we were
going to England."

"We *are* going to England," Cardini responded. "And you can take
that to the bank."

"Cardini speaks!" Spotted Bear declared.

"Hey, kiss my ass, Injun!"

"You kiss *my* ass, wop!"

"Hey, Silverman," I said, "before these two start kissing each other's
asses, tell me something. If we're going to England, then why the hell
are we going north? I thought the North Pole was north and England
was east."

"Look," he said, pointing to a spot on his helmet as if it were a globe.
"New York is here. Okay. England is over here. See? It's east, but the only
way to get there is to go north. That's why they said we were passing
near Greenland. See, Greenland's here. No, wait a minute." He double-
checked the coordinates on his helmet. "Yeah, it's right here."

Chandler looked at him as if he were a nut. Cardini raised his eyebrows and walked away. Averitt said, "Everybody knows that," which
wasn't exactly true, since none of us knew that. Duthie, ever polite,
said, "Okay, that makes sense now." Spotted Bear, who was watching
the global demonstration, said, "Hey, Silverman."

"Huh?"

"I've been wondering why your eyes are so brown, and I think I
finally figured it out."

Silverman looked a little confused. "What are you talking about?
Whataya mean, you know why my eyes are brown?"

"Yeah," Spotted Bear answered. "Because you're completely full of
shit."

That was an old joke. I was surprised Silverman had fallen for it.
"You can go to hell," he said. "You guys don't have to believe me.
What the hell do I care? You'll see, and then you'll thank me."

"Yeah, thank you for the geography class, you stupid little kike,"
Chicago derided.

"I'm not a kike!" Silverman shot back. "I'm a Russian Jew!" For
some reason Silverman—who had a slight speech impediment on top
of his New York accent—pronounced the word "Russian" as if it were
spelled with a *W*, like "Wussian."

Spotted Bear stood up. "Hey, don't be calling the little fella a kike.

He's a— What did you say you were, Silverman?"

"A Wussian Jew."

"He's a Wussian Jew, damnit."

"Hey, Silverman," Chicago said, "why don't you show us Wussia on that globe of yours?"

"You go to hell."

"You go to hell," Chicago mocked Silverman's voice. "Hey, tell me something, Jew boy, is hell somewhere on that globe you got there?"

"No, it's up your ass," responded Silverman, who had been playing along but was now getting upset.

Chandler, like the rest of us, could sense the tensions increasing. But unlike the rest of us, he decided to deal with the situation. He leveled an intense stare at Chicago and spoke. "Leave him alone before I kick your ass off this boat."

From behind we heard First Sergeant Thomas. "What's going on here?"

"Oh, nothing," Duthie answered.

"Always somebody looking to kick an ass or two, that's all," said Spotted Bear. "Ain't that right, Chandler?"

"You ready to fight somebody, Chandler?" Thomas asked.

"Says he can't wait to get into the war," Sam Martin answered.

The first sergeant lit up a cigarette. "I can understand that."

Chandler was still staring at Chicago. "I'll kick your ass right here and now if you like."

"Okay," Thomas instructed. "Save it for the Germans." He looked around at each of us for a minute. "Look, men, this ship is pretty tight quarters. Try not to get in each other's way too much. Go up on deck. Get some fresh air."

"Too damn cold up there," Spotted Bear said.

"It's feeling warmer to me. They say we're passing through some sort of warm southern air current."

"Southern" was the only word I needed to hear. I was up on deck in a flash. It wasn't exactly warm in the "Southern" sense of the word, but it was much more pleasant than it had been before. It was just getting dark, and I could still make out the forms of some of the other ships in the convoy. The full moon was rising over the eastern horizon as the sun was setting to the west, and for a brief time you could see the reflections of both on the water. The seas were calm, and I became fascinated by the wake of the ships as we plowed along. Symmetrical waves split back from the bow of each vessel, little crests of white lining the tops of each as they parted and spread behind the boats. At some point the wakes from different ships would collide. The more dominate ones, the ones from the *Texas,* would overlap the ones from our ship, creating a small upheaval where they met. And that point of intersection followed alongside us as if it were another tiny vessel struggling to keep up.

As the sky darkened and the moon rose, I could make out phosphorus forms glowing in the water beneath us, giving the night a look and a feel almost indescribable. There was something immensely powerful yet immensely tranquil about the sea. I felt helpless and small in the vast expanse of water, yet strangely secure in the man-made machine. I could feel a sense of peace and calm when the sea was smooth, yet I could at the same time sense the deadly force possibly lurking just below the surface as we steamed along. I felt as if I could actually look out and see the edges of the world, though the only thing I could really see in any direction was water. Most of all I felt lonely.

I was on a ship with several thousand men, yet as I gazed out into the night and watched the water fly beneath me, I felt completely isolated. What were Mama and Papa doing right now? What was it like back home tonight? And why can't I be there instead of here?

"That moon is a pretty sight," said Hodges, who had been standing beside me at the railing for the last half hour or so. Several of the guys had been there, but we'd each been alone with our thoughts.

"I can't get over that damn phosphorus," Cardini said, pointing out into the water. "It's spooky as hell."

"I kinda like it," Silverman said quietly.

Sam Martin looked up from the water. "Makes me want to go home."

"Well, the whole damn thing makes me want to go home," Averitt replied.

"Yeah, me, too," I said.

"Me, too," Hutter joined in.

"Hey," Spotted Bear called out, "anybody here *not* wanting to go home? Well, anybody except Silverman? He likes looking at the water."

"It's pretty, ain't it, Silverman?" Averitt gibed.

We all laughed a bit. "You guys can go to hell," Silverman said.

"I think that's where we're going, buddy," replied Duthie. "I think that's exactly where they're shipping us off to."

A few sailors came walking up past us. "Howdy," Spotted Bear said.

"Hi."

"Nice night, huh?"

"It's okay," one of them said.

"How about that beautiful full moon?" Hodges asked.

"What about it?"

"Quite a sight, ain't it?"

"Yeah," said the sailor, looking around at the sky and sea. "Yep, great submarine weather."

The smiles suddenly left our faces, and we peered intently into the night, straining to see any traces of a phantom Nazi sub.

"Come on," Spotted Bear said. "Let's get the hell outta here."

A few of them went belowdecks again, apparently assuming they would be immune from a possible torpedo attack there. Hodges, Duthie, Chandler, Silverman, and I stayed outside and talked until Popeye blew the whistle for bed.

On the afternoon of January 7, 1944, after ten days at sea, we spotted the green southern coast of Ireland off the port side of the ship. According to Silverman's helmet, we should have been in the immediate vicinity of Iceland. In any event, we saw Ireland, but we didn't stop there. Instead we looped back around and headed north to England, a move that made both Silverman and Cardini feel somewhat justified.

We continued north into the Irish Sea, a relatively small parcel of water wedged in between England and Ireland, then headed east until we reached the British port city of Liverpool. As we pulled into the docks, we could see a mass of people waving to us and cheering. A military band was playing Sousa marches. Most of the guys on board ran to the port side to witness the spectacle. It reminded me of the scene when we left New York, except for the part about the band and the fact that I didn't know any of these English people. Then again, I really didn't know any of those people in New York either, but at least they were Americans. Of course, the British were our allies, and we had heard that they were very similar to Americans in lots of ways, except that they talked funny.

It was a great scene as the soldiers on board cheered and waved to the enthusiastic crowd. Some of the guys were so appreciative of the welcome that they started tossing cigarettes to the dockworkers. More specifically, they were tossing the Raleighs. I even threw a few packs myself. Cigarettes were flying off the ship like fireworks, causing a commotion wherever they landed on the docks.

"Look at the poor bastards!" yelled Cardini as the celebrating dockworkers scampered around picking up the Raleighs as if they were made of gold. "You'd think these were the first cigarettes they've seen since the Germans started bombing their asses."

"Times must be pretty bad here," said Silverman.

"Either that or the English smokes are even worse than this shit," added Spotted Bear.

Duthie took a pack from his pocket and threw it as high and far as he could. "That's okay," he said. "I knew these fartsticks would come in handy for something."

First Sergeant Thomas came up next to me at the railing. "Watch this, boys," he said as he launched a pack of Raleighs dockward. The package arched through the air and fell into the water a few feet short of the dock. Two dockworkers jumped in after it.

"Too bad, First Sergeant Thomas," I said. "You have to just throw 'em real hard. They'll get there. I got one all the way over their heads."

"What are you talking about, Sacco? Lookee here!" He tossed another pack high in the air. This time a worker jumped off the dock and caught it in midair before splashing down into the drink. The first sergeant laughed out loud. It didn't take long for half the guys on the boat to start throwing the cigarettes just short of the dock, sending the dockworkers into the water in droves after them.

"Funny as hell," First Sergeant said. "Oh, and by the way, Sacco. Listen, you don't have to call me 'First Sergeant Thomas' every time you address me, okay?"

"Okay, First Sergeant—"

"Just 'Sergeant' or 'Sergeant Thomas' will do. Hell, call me 'Sarge' like everybody else does. That's okay."

"But you said to call—"

"I know. But, hell, we got ourselves a war to win here, so let's just relax and focus on kicking some ass as a unit, okay?"

"Okay . . . Sarge."

"Okay, soldier." He took something from the inside pocket of his coat. "Here, Sacco. Launch these Raleighs just like I taught you."

I did it. A couple of the dockworkers jumped into the water after them. And we laughed our asses off.

We docked at Liverpool, but, much to Cardini's dismay, we didn't disembark there. Instead, after a few hours—during which time some aviation fuel and a few of the other troops were unloaded—we set sail again. This time we traversed the Irish Sea to the northwest, skirting the Isle of Man and the coast of Scotland until we reached Belfast, Northern Ireland.

Here there were no bands, no cheering crowds, no dockworkers flipping into the sea—just a series of trucks waiting to take us to the train station.

7

Tales of the Emerald Isle

Belfast, Northern Ireland
January 11, 1944

The Great Northern Railroad Station's main terminal building in Belfast was cavernous and ornate, with faded decorations dimly glittering on its vaulted ceiling. Shafts of sunlight from the high windows angled around the room, causing the musty air to sparkle within each beam and making the terminal look much more colossal—and dustier—than it probably was. The concrete floors and stone walls echoed the footfalls and conversations of thousands of soldiers until the cadences blended together to become one sound, mixed and muted in a way by the sheer size of the place, yet still rich enough in texture to fill the expansive scene.

Troops were hustling about in a fairly organized fashion, each man carrying his belongings in a duffel bag over his shoulder, each making his way to the trains which would carry all of them on the next leg of their journey. Officers were standing atop benches and small makeshift platforms in an attempt to round up soldiers who were understandably weary and disoriented after a long sea voyage.

All of us in the 92nd were gathered together at one end of the train station, sitting on our duffel bags waiting for instructions on where to go next. Sergeant Thomas was standing on a chair behind us, waving his arms like a traffic cop toward something beyond the scope of our vision. Then, as if by magic, a space opened in the ever-moving sea of humanity, and a small cart, pushed by two Red Cross nurses, emerged, heading in our direction.

"What have we here?" asked Chandler.

"Room service," replied Sarge.

"Would you look at that!" said an excited Duthie as he quickly got to his feet. "That lady has doughnuts!"

"And coffee," one of the nurses added. "Hot coffee and doughnuts, boys. All you want."

We were gathered around the cart in a flash. "Take it easy, soldiers," the other nurse said. "There's plenty to go around. Here, one at a time."

They had a big stack of Army newspapers, called the *Stars and Stripes,* for us to take if we wanted to read and catch up on the war. They were also giving out free cigarettes—Raleighs—which most of us declined.

"I'll take mine black," requested Averitt. "And a doughnut. Can I get two?"

"Here you go, soldier," one of the nurses replied as she handed Averitt a cup of joe and two doughnuts. "You can have as much as you like. That's what we're here for. Here, you want a pack of cigarettes to go with that?"

"No thanks."

"Get the hell away from there, Averitt," admonished Cardini as he pushed up to the front.

"Yeah, Averitt. Just get your damn coffee and move out of the way," a voice from the back called out.

All the guys feasted that day, thanks to the Red Cross. There was nothing particularly sumptuous about the coffee or doughnuts, but after being on that boat for so long, we felt like we were in something of a culinary heaven. The best part was that this was good old 100 percent grade-A American coffee, not the seawater-and-piss variety they were serving on the boat. I got mine black, and lots of it. I must have eaten ten doughnuts and would have consumed about a dozen more if Sarge hadn't called us to head to our train. Most of the guys stuffed a doughnut or two into their pockets as we gathered up our belongings and moved out. I wrapped four in a napkin and packed them into my hat. I had to smash them a bit, but they were gonna get all smashed up in my stomach anyway, so what did I care?

The Irish trains weren't like American ones I'd been on. These had seats facing each other with tables in between, so half of us were always riding backward. That felt a little strange, especially when we first pulled out of the station, but not as bad as rocking around on the ocean for ten days. All in all, it was good to be on dry land again. And once we cleared the city of Belfast, it was obvious that this particular land was one of great beauty.

"People here speak with a brogue," Cardini informed us as we watched the Irish countryside roll past us.

"A brogue?" I asked. I had never heard that word.

"Somebody said they've got good whiskey," said Spotted Bear.

"Yeah," Silverman replied. "That's what they call it—an 'Irish brogue.' "

"A brogue," I pondered.

"It's like an accent," Averitt clarified.

"Why don't they call it an accent?" I asked.

"Because it's an Irish brogue," Cardini insisted. I didn't think his answer gave me a lot of information, but I was willing to play along. I was personally looking forward to hearing the Irish talk. Up until this point, the strangest accent I'd ever heard was coming out of Silverman, so I was curious as to how these folks could top that.

"I hear they got good-looking girls here," said Averitt. "All red-haired, blue-eyed, and freckled."

"All of 'em?" Chandler asked.

"Most all of 'em," Averitt answered.

"Freckled?" Spotted Bear asked. "And talking with those brogues?"

"And big-breasted," Cardini added authoritatively as he simulated a rather buxom shape with his hands.

"How do you know so much about the Irish?" I asked Cardini.

"Whataya mean? From the priest at church!"

"What the hell? The priest told you that Irish girls have big titties?" Spotted Bear asked.

"Oh, hell no," Cardini answered. "What's wrong with you? Priests don't talk about things like that! He just talks about what things are like in Ireland."

"How does he know?" I asked.

"How the hell do you think he knows?" Cardini was surprised I would ask. "He's *from* Ireland. All the priests are from Ireland. Where do you think they come from? Ain't your priest from Ireland?"

"No," I answered. "Father Canapa's from Italy. But now that I think about it, that priest back in basic, he was from Ireland or someplace. I think I remember him talking with a brogue."

"See?"

"Hey, Cardini," Spotted Bear interrupted, "if it's okay, let's don't talk about the priest anymore. Let's talk about the girls with the big knockers."

"Good idea," Chicago added.

"Hey, how long you think we're gonna be here?" Duthie asked. "Silverman, what's the name of that place we're going?"

Spotted Bear was quick to respond. "Hey, Duthie, unless the answer is 'titties,' don't ask a question."

"And all the girls here, they call 'em 'lassies' and 'colleens,' " Averitt continued.

"Lassies and collies?" asked an incredulous Chicago. "Like the dog?"

"No, not collies," Averitt answered.

"Coll*ee*ns," Cardini stressed. "Is it possible you're that damn dumb? 'Colleen' is what they call a girl. They call 'em 'colleens' and 'lassies.' "

"Dumb as my ass," Averitt sighed.

Chandler got a serious look on his face and leaned across the table. "Well, actually, *all* of you are dumber than Averitt's ass—and that includes you, Averitt—because it's the Scottish who call a girl a 'lassie.' The Irish call them 'lass,' if you wanna get real precise about it. And a collie is a Scottish breed of dog. And 'colleen' is from an old Irish word that means 'girl.' "

"That's what I said," Cardini stated weakly in his own defense. The rest of us had stopped talking and were staring at Chandler.

"Uh . . . ," he stammered. "I was in pr—Uh . . . I knew a guy back in the . . . back home who was from Ireland. And he wasn't exactly a priest."

Things were quiet for a few seconds. Then Silverman spoke. "Lurgan."

"Huh?" asked Spotted Bear.

"Lurgan. That's the name of the place we're going."

"Oh," Spotted Bear responded. "I'm still thinking about the concept of Averitt being dumber than his own ass."

Lurgan, Northern Ireland
January 1944

Lurgan was a small town about forty miles southwest of Belfast. About fifteen hundred people called it home, but it seemed to be populated mainly by sheep, goats, and chickens. Set among the scenic hills of Northern Ireland, it was, I suppose, located just far enough away from the big city to keep us out of trouble. The Army—for reasons known only to the Army—deemed it the perfect place for us to be. So, with absolutely no fanfare whatsoever, the 92nd Signal Battalion rolled into town.

The journey took us only about an hour or so, but within that short span of time, a mist had rolled in and the sky had grown dark. Night had settled quickly in this part of the world. In fact, by the time we disembarked from the train, it was pitch-black. The strange part was that it was three-thirty in the afternoon.

"Is this thing broke?" Averitt asked as he tapped his watch. "It's saying 15:30."

"No, you're right. It's 15:30," Sarge informed him.

"Kinda early to be so dark, ain't it, Sarge?" asked Spotted Bear.

"High northern latitude," the Sergeant responded. "Only a few hours' daylight. Gotta make 'em count."

Silverman took this as his cue to remove his helmet and point out our global position. I started laughing. Sarge wasn't so amused. "Put that helmet back on, soldier," he snapped.

"Yeah," Chicago added. "Before somebody crams it up your ass."

Chandler, who had been happily minding his own business while munching on one of the many doughnuts from his stash, stopped chewing, set down his duffel bag, and looked directly at Chicago in that scary, I-might-have-to-rip-your-face-off way.

"Okay, let's move out," Sergeant Thomas ordered. "Stick together."

And with that we began the short march south from the train station to our campsite. There was a blackout policy here, so we had to be especially careful of tripping over each other or falling off the road altogether, since we couldn't see where we were going. Most of the guys were looking down at the ground, trying to discern where to take the next step. A few were looking up and following the silhouettes of the trees against the starlit sky. Either way, it wasn't easy. Slowly our eyes adjusted to the lack of light as we cautiously stumbled our way through the moonless and cold Irish night.

Our camp was at the edge of town and was in itself a small village of about seventy-five quonset huts, each able to accommodate twelve to fifteen men. There were two windows at each end of the hut, and they were covered with blankets because of the blackout. The beds were constructed of a piece of plywood resting on two short sawhorses, just high enough to keep us off the floor. A large bag filled with straw, presumably the Irish version of a mattress, covered the plywood. None of us were particularly enthused about sleeping on this stuff, but then again, it was better than sleeping on the ground. Actually, the mattress was pretty soft, but in an uncomfortable, scratchy kind of way. I suppose when the Irish said it was time to hit the hay, they really meant it.

Straw beds or not, we were accustomed to getting out of the rack before sunrise. Of course, in Lurgan in January, the sun didn't rise until 10:00 A.M. But that didn't mean we were allowed to sleep late; it just meant that we spent a great deal of time bumping around in the dark. To make matters worse, the somewhat lazy sun would set again at 3:00 P.M., plunging us back into what seemed like a perpetual nighttime. Operating in a blackout made even the simplest of tasks difficult, especially when there was no moon to illuminate our surroundings. And apparently the Army had ordered the moon out of the sky, because I don't remember seeing it for the next two months.

Not only was it always dark, it was also cold and damp, a combination that seemed to activate everyone's bladder. When we weren't peeing, we were alternating between a few days' KP duty and a few days' guard duty, then back to KP, then back to guard duty. Other than that, there wasn't really anything for us to do around camp other than complain. And after a while that got old, too. Then we'd just go back to peeing, quite possibly out of sheer boredom.

The food served up in the mess tent was almost as bad as the caca they had been serving on the boat. Despite the proliferation of chickens running around Lurgan and the associated availability of authentic eggs, the Army had developed a food substitute called "powdered eggs," which they proudly served every morning and which, despite the cooks' best efforts, tasted pretty awful. We had to eat, but who wanted to eat this crap?

Fortunately, our training included going out into Lurgan and some of the surrounding villages to practice installing phone lines. On those days we were allowed to break for lunch and eat with the civilians at local restaurants. We had been told there were food shortages in the United Kingdom, but we somehow always seemed to find beefsteaks and potatoes.

Lurgan itself consisted of one main street dotted with shops, bakeries, other businesses, and what seemed like an inordinate number of bars—or pubs, as the Irish called them. There was a church, a school, a market, a post office, a doctor, and all the other normal things one would expect to find in a small Irish town, including sheep. Without a doubt, the most popular establishment among the GIs was Lurgan's lone fish 'n' chips restaurant, which we frequented as often as possible. When we couldn't get a pass to go into town ourselves, we'd pay the local kids to deliver fish 'n' chips to the fence behind our hut.

The outlying areas consisted of beautiful green hills and meadows, dotted here and there with tiny villages and towns. Spaced across the landscape were picturesque cottages with thatched roofs and stone chimneys churning warm smoke up into the cold, damp air. Rock walls, two or three feet high but stretching for miles, cut white lines through the fields, while clouds and mist settled between trees throughout the countryside, making the panorama seem more like a painting than a real place.

We were the only American troops here, for which I'm sure the people of Lurgan were most thankful. On the whole, they were a jovial, friendly group, quite welcoming of the nine hundred soldiers of the 92nd Signal Battalion who had invaded their town. Everybody spoke with the Irish brogue, some so thick I couldn't understand them. But those were usually the old men, so I really didn't care what the hell they were saying. We would just nod and smile and keep going about our business. They were nice to us, we were nice to them. Everybody was nice.

All in all, Lurgan wasn't what you'd call exciting. I suppose it had a certain old-world charm, but when you're nineteen years old and a long ways from home, you're not exactly looking for charm. You're looking for chicks. In our case we were looking for chicks and decent food.

Our plan the evening of February 6, 1944, as it had been on so many evenings before, was to visit every bar in town. The logistics of this feat weren't quite as complex as they may seem, since the main street into town—the street with all the bars on it—led right up to the gate of our camp. Our custom was to imbibe at each establishment all the way up one side of the street and then just drink our way back through each one on the other side. On any other night, the big challenge would have been simply to remain sober enough to walk upright back into camp. But on this particular Saturday, we had been invited to attend a special dance the town was having in our honor. This was, of course, after the drinking.

"Why are we waiting on Silverman?" Chicago asked. He was irritated. "Come on, let's go."

"Yeah, come on, let's leave his scrawny ass," Tex suggested.

"Just give him a minute," I said. "He'll be here."

Cardini was looking back into the camp. "Here he comes," he said. It was dark, so it was hard to see much of anything. "Come on, Silverman! Shake a leg!"

"Hey, I'm coming! You guys hold up!" the voice of the runner called out. It wasn't Silverman. It was Wilkins.

"Oh, shit," sighed Cardini. It was too late. Wilkins was upon us.

"Okay, guys," he said. "I'm ready. I had to take a shit."

"We didn't need to know that, Wilkins," said Averitt.

"Don't shake his hand," instructed Sam Martin. "It might be the one he wiped his ass with."

"Not planning on it," replied Averitt.

"What makes you think he wiped his ass?" Spotted Bear asked.

"Wilkins," said Duthie as we began walking away, "you realize you're not required to tell us about every shit you take."

"Yeah, I know," Wilkins continued. "I been taking shits all day . . . since . . . all day. One shit after the other."

Cardini had heard enough. "Somebody talk about something else," he said.

"You think I like it, Cardini?" Wilkins asked, as if his feelings were hurt. "I wish I could stop shitting. It would make my ass feel a lot better."

"You know what?" I said. "There's an old saying: 'Make a wish in one hand and take a shit in the other and see which one comes true.'"

"In that case," added Sam Martin, "make sure not to shake either one of his hands."

We were halfway to the first bar when we heard someone running toward us from the direction of camp. "Hey, you guys, wait up! Hold on a minute!" This time it was Silverman.

"Look at this," Spotted Bear noted. "This really *is* a rushin' Jew."

"Where the hell have you been?" Cardini asked. "You been taking a dump, too?"

"Huh?" Silverman looked confused. "No . . ."

"Come on, don't worry about it," Duthie said.

"I ran into a captain who said my belt buckle wasn't polished, so I had to—"

"Okay, Silverman," Chicago interrupted. "We get the picture."

"You know where I been all day?" Wilkins asked Silverman.

"I give up," Silverman answered.

Chandler gave Wilkins a look and said, "Don't start again."

"He means don't say 'taking a shit,' " Averitt quickly added.

"On the crapper taking a shit," Wilkins proclaimed, seemingly oblivious to the fact that Chandler could have killed him on the spot and the rest of us would have helped bury the body.

Spotted Bear shook his head. "Never met anybody so proud of laying cable," he said as we walked along.

As we entered the first bar, I spotted Hodges sitting with some of his buddies. I went over to him. "Hey, guess what?" I whispered.

"What?"

"I think Wilkins got some of the coffee."

"No shit?"

"Lots of shit," I said. "Been shitting all day."

"What're you talking about, Sacco?" one of the guys at his table asked.

I looked around at Wilkins, who was getting settled across the room with our group. "You tell 'em," I said.

Earlier that morning Hodges and I had been on KP duty. As I was reaching for a can of lard up on one of the high shelves, I knocked a few items onto the floor. One of the items, a big bar of unwrapped GI soap, slipped off the end of the shelf. I heard a splash.

"Did I just see what I thought I saw?" a wide-eyed Hodges quietly intoned.

"What was it?" I asked. "Soap?"

"Uh-huh."

"Where did it go?"

"Where the hell do you think it went? It fell in the coffee," Hodges said as he wiped off his hands and picked up a large wooden ladle. "Shit! We better get that outta there before it melts."

He had climbed above the large vat of brewing coffee and was stirring around with the ladle trying to get the soap out. Several times he twisted his face and said, "Hmmm . . . I got it"—only to have it slip off the side and fall to the bottom again.

A colonel got his attention. "Soldier!"

Hodges looked around. "Yes, sir?"

"Here," the colonel said, handing Hodges his cup. "Give me a cup of coffee."

"But—"

"A cup of coffee, soldier! And I don't have all day!"

"Yes, sir," Hodges said as he dipped the officer a cup of soapy joe. The colonel put it on the tray next to his powdered eggs and walked away. He was, in all likelihood, shortly indisposed. Probably blamed it on the eggs.

We were on duty for about an hour longer, so we made sure that nobody else drank the coffee—at least nobody we liked. We never actually saw Wilkins, although I'm certain we would have been happy to give him a double portion if he'd wanted some.

In any event, as we now made our traditional round of the bars, Wilkins started looking more and more sick. All the while he was drinking Guinness, which was something of a laxative in itself. By the time we got about halfway down the street on the way back to camp, he was doing some major squirming.

Cardini pushed another beer in front of him. "Here, drink this. It'll make you feel better."

"You guys leave me alone. I feel like shit."

"And you smell like shit, too," Chicago said. "Did I ever tell you that, Wilkins?"

"And he looks like shit, too," added Spotted Bear.

"And likes to talk about shit," added Averitt.

"Hey, what time are we going to the dance?" Wilkins asked.

"Right now," Cardini answered.

"We've got three more bars to go to. Then we're heading over there," Averitt corrected.

Wilkins had been fidgeting around more and more as we talked. "Look, I've got to run to camp for a few minutes. Where will you guys be?"

"We'll be gone," Spotted Bear responded.

"No, wait on me," Wilkins said as he squirmed his way to the door.

"Yeah, like hell," I said.

"Come on, guys, I'll be right back."

All the guys were looking at him and laughing. "Awww, hurry up and go take your shit!" said an irritated Tex.

Wilkins mumbled something as he bolted out the door, but I couldn't understand what it was. All I know is that he hopped up on one leg and grabbed his belt before proceeding to dance down the street as if he had a stick shoved up his ass.

"Looks kinda like a bird, don't he?" Spotted Bear noted.

He had made it only about a quarter of a block when he suddenly stopped, bent over, lifted a leg, partially sat, twisted his hips around, clutched his stomach, and then crouched down into an all-out squat. We ran to the window of the bar, straining to see what he was doing. He remained in that position for a few minutes. Then, slowly, he straightened himself up and began walking stiff-legged back to camp. The back of his Class A dress uniform was, let us say, no longer regulation Army green.

From that day forward, Wilkins assumed the name of "Bird Turd." No one ever—and I mean *ever*—referred to him by any other name for the rest of the war.

Needless to say, Bird Turd didn't make it to the dance that night. But the rest of us did. It was held at the town hall, at the opposite end of the main street from our camp, near the church. We were the only American troops in Lurgan, and the townspeople went out of their way to put on a good dance for us. They had red, white, and blue banners hanging all over the place, a band playing American music, lots of food, and plenty to drink. Of course, after making our rounds of the bars, drinks were not exactly what we needed, with the possible exception of coffee.

It seemed like every girl in town showed up. Actually, they must have imported some from other towns, because all these couldn't have been locals. Even with several hundred men of the 92nd in attendance, there were still many more girls than GIs. Maybe it was the appeal of the handsome foreign soldier in their country to save the day, maybe it was the dress uniform, or maybe it was just the liquor—but whatever it was, it was working, making it almost impossible for us not to capture the attention and admiration of the girls of Lurgan.

By mid-February the days were finally starting to get a little longer. It felt good to get a few extra hours of daylight, but the fact of the matter was that the sun still wasn't up until eight in the morning and it was pitch-black by six P.M., so most of what we did continued to be done in the dark. After a while we started to feel like we were living in a cave, and it began to wear on us.

As might be expected, the blackout policy made the darkness seem even darker. We had blankets covering the windows, so that if we wanted to, we could keep the lights on inside our huts without giving away our position to enemy reconnaissance aircraft. Then again, while we were in the hut, there was nothing to do but look at each other. So most of the time we'd turn off the lights and go to sleep.

And going to sleep wasn't always the easiest thing in the world. First of all, our systems were out of kilter because of the prolonged darkness. Then Silverman was usually wheezing and sneezing, because he was allergic to the straw in the mattresses. And, to make matters worse, Chandler and Chicago took turns farting all night—and I mean loud ones that would resonate off the metal hut and rattle around for a few seconds before they died down. Averitt would say, "This is what they mean by gas warfare."

I was drifting off one night when I thought I heard someone whispering my name.

"Hey, Sacco . . . Sacco." It was Cardini in the next bed over from me.

"Huh?"

"Come here for a minute."

"What?"

"Come here. For just one minute."

I could tell that he had a flashlight and was looking at something under his blanket. *What the hell is this?* I thought as I rolled off my bag of hay.

"Here," he said, handing me the flashlight. "Hold this."

"What are you doing?"

"Just hold it right here, like this," he responded. "I need to see . . . right here."

I peered under the blanket. "What are you looking at?"

"Just hold the damn light," he answered. "And hold it still. I'm trying to pick these damn crabs, and it's taking both hands because I have to—"

I turned off the flashlight and dropped it.

"Hey, what are you doing?" he asked.

I hopped back into my bed, thought for a second, and then answered, "I'm trying to figure out who's the bigger dumb-ass, me or you."

"Oh, come on, Sacco. What's the big deal?"

"Hey," I said, "when the sun comes up, you can sit outside next to a tree and pick your damn crotch all day. How about that?"

"Come on, Sacco. It itches."

"Trying to sleep over here, Cardini," I said as I pulled the covers over my head.

The next morning, just after reveille, I heard Cardini once again calling my name. "Go pick your own crotch," I said to him, much to the amused bewilderment of Averitt and Duthie, who happened to be standing nearby.

"No, listen, Sacco," he said. "You answer for me today. I'm gonna sleep a little more."

"Why don't you get Bird Turd to answer for you like you normally do?" Duthie asked.

"Come on, Sacco," Cardini pleaded. "You owe me."

I don't know how he figured I owed him anything—and I don't know why I decided to play along with any of his scams—but I said, "Okay, but just this once."

A few minutes later, at roll call, I did my best to emulate Cardini's voice under cover of the early-morning darkness.

"Sacco?"

"Yes, Sergeant?"

"Why are you answering for Cardini?"

"Uh . . . that wasn't me, First Sergeant Thomas."

"Sacco!"

"Yes, Sergeant?"

"Go get Cardini and tell him to get his sleepy ass out here right now."

"Yes, Sergeant."

As I was leaving formation to go back in the hut, I heard Averitt say, "Be careful there, buddy. Cardini might be a little crabby."

There were basically two types of passes, Lurgan passes and Belfast passes. Here's the way it worked: With both passes we had to report back to camp by 11:00 P.M., but the Belfast passes allowed us to leave camp a little earlier, at 3:00 P.M., whereas the Lurgan passes didn't let us leave until 5:00 P.M. The extra time was due to the train schedules from Lurgan to Belfast and back.

Some of the guys would get Lurgan passes, then run off to Belfast. It was possible to do so without getting in trouble *if* you worked the trains just right and there were no delays. If the train from Belfast was late, you could get the conductor to write an excuse on your pass and you wouldn't get into trouble—that is, if you had a Belfast pass in the first place. Otherwise you had no excuse, and you'd end up working double duty on KP.

One sunny, mid-March Saturday afternoon, Cardini bounded into the hut sporting eight Belfast passes. "Get ready, boys!" he proclaimed. "We're gonna have one helluva night on the town!"

"Where did you get those?" asked Silverman.

"Where do you think?" Cardini replied as he began handing out the passes. "From the captain."

"Yeah, well, I figured that," Silverman responded. "It's just that I didn't think they were giving out Belfast passes for the next two weeks."

"That's the problem, Silverman," Cardini shot back. "You don't think. I got the passes, don't I?"

"See," Spotted Bear said as he took one of the passes and slipped it in his pocket, "I told you if we put up with enough of Cardini's bullshit, it would be worth it one day."

"You can thank me later, Injun," Cardini said.

"I'm not thinking about you at all, my Eye-talian friend," said Spotted Bear. "I'm just gonna concentrate on relaxing and having a good time. And maybe I'll buy you a drink or a hooker or something."

"What about you, Sacco?" Cardini asked. "You gonna get you some hot bread tonight?"

"Bread?" I didn't know what he was talking about. "I guess it depends on where we eat dinner."

The other guys started laughing. "I don't think he means that type of bread," said Averitt.

I had heard people in movies say "bread" when they meant money, and of course I'd heard people say "bread" when they meant bread, but I still didn't know what Cardini was getting at. After all, we weren't going into town to rob a bank or a bakery or anything.

"I'm sure as hell gonna get some," said a smiling Chandler. "And the hotter the better."

"You really don't know what he's talking about, do you?" Averitt asked.

I shrugged my shoulders. "I don't see what the big deal is. We can get all the bread we want in the mess tent."

The other guys were laughing. Cardini spoke up. "You see a real shapely-looking piece of Irish ass walking down the street in Belfast— hot cross buns, my friend. Get 'em while the getting's good."

"Yeah, well, I'd rather just say 'a piece of ass' like we do back home and not bring food into it," I said.

"Hmmm," Averitt added. "Apparently Sacco don't like to mix his food with his ass." Averitt, of course, was an expert when it came to the ass.

"Time's a-wasting, boys," Cardini said, brandishing his pass above his head. "And our adventure awaits."

Cardini, Averitt, Duthie, Spotted Bear, Chicago, Chandler, Silverman, and I put on our dress uniforms, polished our shoes, and headed for the big city. Belfast was a nice town, and we had a good time there. It seemed to be filled with American GIs, all of us—with the exception of the faithfully married soldiers, such as Duthie—looking for Irishwomen. After dinner (at which we did get the kind of bread you eat) we set sail on a night of drinking at every pub we could find.

We were walking down one of the main streets when we saw a commotion coming toward us. Ten or fifteen children were laughing and talking excitedly. In the middle of them, standing at least seven feet tall, was a black man in a striped zoot suit slowly making his way down the sidewalk as the children orbited around him.

"What's going on?" Chicago asked.

"Would you look at that?" Averitt said.

The children were mesmerized by the man and asked him dozens of questions in rapid succession. "How did you get so dark?" "How did you get so tall?" "Are you from Africa?" "Do you have a camel?" "Are you a giant?" "Can you jump over a building?"

The man, for his part, gently laughed and answered each question as best he could while he walked along. As he passed us with his entourage, he looked over and gave us a salute. "Hello, American soldiers," he said with an African accent.

The children, who normally would have been impressed by our uniforms, looked around at us briefly, then turned their attention back to their giant. Hell, we were looking at him, too. He made all of us look like shrimps.

"Ain't that one helluva sight to see?" Cardini said, shaking his head. "I wonder where he's from."

"Something tells me he's not from around these parts," Averitt said.

"Gee, what makes you say that, Averitt?" Silverman asked.

"Easy one. The folks from around here don't wear zoot suits," Averitt answered.

"That's not a zoot suit, Averitt," Silverman corrected.

"I know a zoot suit when I see one, midget," Averitt shot back.

"You don't know shit," Silverman said. "That's just a suit with stripes. A zoot suit has these—"

"Can you two shut the hell up for a couple of minutes?" I asked.

Just before the giant and his entourage disappeared around the corner at the end of the block, he stopped, looked back at us, smiled, and waved.

"Ain't that the damnedest thing you ever saw?" Chandler asked.

"Impressive," Chicago responded. "Impressive."

Spotted Bear was waving his hand. "Ah, I've seen bigger niggers than that. They grow 'em big down in Texas."

"You've seen *what*, soldier?" a voice bellowed from behind us. We turned to find three American MPs. The one asking the question was black. "*What* have you seen, soldier?"

Spotted Bear started to point back in the direction of the zoot-suited giant. Silverman was quick to cover for him. "He's drunk. That's all."

"Let me see some passes and some ID," the MP ordered.

After carefully studying each ID and pass, the black one said,

"Okay, now, General Eisenhower says that everybody out here is equal, so—"

"Yes, sir," Spotted Bear said.

"Don't call me 'sir,'" the MP countered.

One of the white MPs asked, "You see an officer around here, soldier?"

"No, sir," Spotted Bear replied.

"Drunk *and* stupid," Silverman said.

"So General Eisenhower says everybody here is equal, so that's enough of the prejudiced talk against colored people," continued the black MP. "Is that understood?"

Spotted Bear said, "Yes, sir," but the rest of us just said, "Yep" or "Understood" or "Okay."

"Okay, then," said one of the white MPs. "Go on."

"Don't want to see any of you again tonight," said the other. "So try to keep your heads out of your asses and your drunk friend's mouth shut."

We knew that Spotted Bear, if given the chance, would have said something else and gotten all of us in trouble, so we moved him down the street as quickly as we could. As we were walking away, Chandler slapped him across the back of the head. "Way to go, idiot," he said.

Fortunately, the MPs had let us go on about our business. Some of the characters we encountered on the streets of Belfast weren't as congenial. In fact, we had just exited a bar and were walking with a few Irish gals when we came upon a group of about eight teenage boys. They had been doing some serious drinking and were looking for a fight. We'd done some drinking ourselves, but not that much, with the possible exception of Spotted Bear, who always seemed to do a little more than his share.

The Irish boys began trying to shove us a bit, but we stepped aside to let them pass. Finally one of them said, "You Yanks are not so tough! Why don't you go back home and leave our women alone?" Another one took a swing at Chandler. It puzzled me as to why anybody would ever try to take a swing at Chandler, since he was by far the biggest and most ferocious-looking one of us.

Chandler simply caught the boy's fist in midflight and twisted his arm behind his back. The Irish lad immediately crumpled to the ground and started crying.

"Leave him be, you Yank son of a bitch!" one of the others yelled.

"Leave him be before we beat the hell out of all of yas," another said, putting up his fists like an angry leprechaun.

The excitement of the moment was not lost on the girls, who began a spirited tongue-lashing of the intoxicated boys. "Why don't yas just go along home and fight against yourselves if you're looking that much for a fight?"

"Go along minding your own business unless you really want to get hurt," another said, almost daring them to mix it up with us.

Another spoke up, "You're a bunch of little horses' asses, you are! I'm supposin' you must be from England, you band of unruly, drunken hooligans!"

"From *England?*" one of boys protested in a rather high-pitched tone. "And look at yourself, cavorting around with the Yanks like they . . . like they . . . Just look at yourself, would you?"

"You got no business being with these Yanks," another of the boys said. "What are ya? Whores?"

"Hey, watch the way you talk to these ladies," Chicago warned.

"Shut your mouth, Yank!" the boy snapped back.

"Or what?" Chicago was getting ready to whip some ass.

"We can be with whomever we like, so run along home," a girl responded, "before they decide to kill the whole lot of you."

Chandler still had the one kid on the ground, holding him there for his own safety, since he seemed to be the drunkest and most belligerent.

"Look, we don't want any trouble," Averitt said.

"Fight like men!" said the one on the ground.

"We didn't come over here to fight you," I said.

"That's because you're scared," one of the boys answered defiantly. He looked at the girls. "They're scared to fight us," he repeated.

The girls were laughing. "Ah . . . ," one of them said. "These here are real soldiers, laddies. They could kill all of yas in one instant and never think about it again."

"Maybe I'll just kill this one," Chandler said as he tightened his grip and pushed the boy's face into the pavement. We couldn't tell if he was kidding or not. Apparently neither could the Irish boys, who stopped and stared wide-eyed at him.

One of the Irish girls spoke. "Go ahead and kill him. It'll teach him a lesson."

Spotted Bear, himself somewhat loaded and therefore finding the whole scenario humorous, kept laughing out loud throughout the exchange.

"What's so funny?" one of the Irish boys asked.

"He's laughing because he knows we can take him and his Yank friends in a fight," another answered.

"Don't waste your time with that one," one of the other boys instructed. "He's not even a real American. He looks like he must be a Mexican or a Negro or something."

Spotted Bear stopped laughing.

Duthie, being the voice of reason, intervened. "Look, we don't want to hurt you boys."

"We understand that you're Irish and you like to fight and all, but we're trained to kill people," I added. "That's what we're trained to do. And we didn't come over here to fight you."

" 'Cause if we fight you, we'll beat the hell out of you right now," Chicago warned.

"And we'll put a real bad ass whoopin' on you," Averitt added. "Real bad. Right now."

Duthie continued, "So just turn around and keep walking, and we won't have any trouble."

The boys were quiet. The one on the ground stopped struggling under Chandler's hand.

"Whatcha think, buddy?" Chandler asked his captive.

There was a hesitation. "Okay," the boy finally said.

"Okay, then," Chandler said as he released the lad. The boy jumped to his feet and brushed himself off.

"Go home and grow up," one of the girls said.

"And on your way there, stop by the market, buy yourself a pound of butter, and slide yourselves off the face of the earth," another added. "What a disgraceful bunch of hooligans!"

It seemed to me that the girls rather enjoyed getting the boys' ire up and were disappointed that the whole incident hadn't erupted into an all-out brawl. The boys, for their part, hurled a few more insults at us but moved along, apparently deciding to look elsewhere for their nightly fisticuffs. All in all, I thought we had handled it pretty well.

"Mexican?" Spotted Bear slurred. "That boy called me a Mexican."

"He also called you a Negro," Chicago reminded him.

"Come on, just forget it," Silverman said.

"I'm not no damn Mexican," Spotted Bear insisted. "And I sure as hell ain't a Negro."

"Nobody cares," said Chandler.

"Come on, Pedro," Averitt wisecracked. "Let's get some coffee into you."

It was right about then that we saw a familiar figure coming out of one of the clubs. "Well, look who's here," Sergeant Thomas said, genuinely surprised to see us.

"What are you doing here, Sarge?" asked Chicago.

"The question is, what are *you* doing here?" Sarge returned.

"Whataya mean?" Chicago answered. "Just getting in some R and R."

"Where did you get the passes?" Sarge asked, getting more serious.

"From the captain," Silverman responded.

Sarge looked puzzled. "From the captain, huh? Well, Captain English reported some passes stolen this afternoon. Eight of them. Asked me if I had any idea where they might have disappeared to."

There was silence as each of us swallowed hard and looked in Cardini's direction.

"Cardini!" Sarge bellowed, as the culprit did his best to become invisible. "Cardini!"

"Uh, yeah Sergeant Thomas?" he finally responded. The rest of us stepped as far out of the way as possible.

"You know anything about these passes, Cardini?"

"Uh . . . no."

"You don't know how these men got these passes?"

"No, Sarge, I don't."

Sarge had heard enough. "Okay, boys, here's the way it is: All of you be on the next train to Lurgan."

"But—" Silverman began.

"Cardini, you're confined to barracks."

"But, Sarge . . . ," Cardini pleaded. "I didn't—"

"The rest of you," Sarge interrupted—"shut up, Cardini—the rest of you don't have a problem. You"—he was pointing at Cardini—"meet me in the captain's office, 0800."

There was a short but uncomfortable silence as we stood there waiting, unsure of what to do next. "Let's go, boys," Sarge said. "Back to the camp, and I mean double time!"

Cardini had some hell to pay for the next few days, with nonstop KP, latrine, and guard duty. The rest of us were let off the hook, having enjoyed a free night on the town at Cardini's expense and then receiving the unexpected pleasure of seeing him get busted.

Life in Lurgan became more bearable as winter turned to spring and the days became longer. Problem was, they started becoming *too* long. By the beginning of May, the sun was rising at 3:00 A.M. and not setting until 10:00 P.M. At that point we were using the blankets over our windows to keep the light out, not in.

The monotony of waiting around for the eventual call to action was becoming rather numbing. Perhaps that's the way the Army had planned it—get us so bored out of our minds that we would look forward to combat. If so, it was working. There were rumors of an impending invasion, of where and when it would be, of how many men and what types of weapons would be sent. But they were all rumors, some started by people like Cardini who didn't know anything, some started by military brass to throw people like Cardini off track. Truth is, we didn't know what was to happen, a fact that merely served to magnify the intensity of the boredom.

The only real excitement came one bright Saturday morning in mid-May, when General George S. Patton made a brief visit to Lurgan to inspect the troops. We stood at attention in a small field within the confines of the campsite as the great general walked up and down the rows, nodding, scowling, grunting, squinting, seemingly deep in thought yet attentive to his charge. Each of us stood a little straighter as he passed, hoping nothing on our uniforms or in our countenances would be amiss.

I tried not to breathe as he approached. He glanced at me and quickly moved on. When he reached Sergeant Thomas, he looked back at us and said, "Congratulations, First Sergeant. You've done a fine job with these men—one helluva job. I'll be happy and proud to lead them into battle." He said it loud enough for the rest of us to hear.

He walked briskly to the front and faced our ranks. "I have it on good authority that you men of the 92nd Signal Battalion are ready for combat. You have received excellent training, and each of you has passed with flying colors. You look damn fine, I'll tell you that right now, like you're ready to kick some Nazi son-of-a-bitching ass from here to Berlin and back. It will be an honor to lead you brave men into battle. I'll be with you every step of the way, in every battle as we crush the head of the enemy. God bless you men, each and every one of you."

And with that he turned and walked away, his regalia gleaming, his entourage in hot pursuit. We stood at attention until his jeep left the marshaling area. The visit had lasted all of twenty minutes, but it had made a powerful impression on us all.

We packed our bags and left Lurgan on the morning of May 23, 1944, setting sail from the port of Larne, Northern Ireland, across the Irish Sea to Stranraer, Scotland, where we disembarked and were whisked away to a nearby British camp for one of the most pathetic meals any of us had ever tasted. Several theories emerged as to what we were consuming, but we tried to be polite, since the English soldiers in attendance actually seemed to be enjoying it.

At 1800 we boarded a troop train and headed south through Scotland. It was a beautiful country of quaint villages and rolling hills, much like Ireland. The ride was uneventful until we rounded a bend just outside the small town of Carlisle, where several of the guys spotted something on the right side of the train.

"What the hell is that?" Hutter asked.

"Damn," sighed Averitt.

"Gimme my rifle," said Spotted Bear as he stood and lowered the window.

"What's going on?" Silverman asked, jumping up from his seat on the other side of the train and peering out the window. "What the . . . ?"

By this time most of us were on one side of the train, looking out at the vast hillside slowly passing. There, almost completely covering it, were thousands and thousands of huge rabbits, hopping over each other, eating grass, staring at us. Spotted Bear had his carbine resting on the window, pointed out at the immense assemblage. The sheer magnitude of critters eliminated the need for any type of marksmanship; a blind shot would probably hit two or three. He nonetheless seemed to be carefully aiming, looking through his scope, then looking up, then looking through his scope again.

Before he could pull the trigger, all the windows were down and all the rifles on board were pointing toward the hill. The next sound could very well have been a volley of gunfire from the train onto the unsuspecting hoppers had not Sergeant Thomas and a couple of lieutenants burst into our car with orders to stand down our weapons.

Not a single shot was fired. We lifted our heads from our sights and, with rifles still trained on the hillside, watched the rabbits as they stared back and happily hopped along, oblivious to the closeness of their sudden and violent demise. The massacre on the hill had been averted. The bunnies would live to see another day.

"Shit," Spotted Bear said under his breath. He seemed genuinely disappointed.

"Sit your asses down right now," Sarge barked out. "You wanna kill something? Kill Germans. And for your information, the krauts will have guns, and they'll shoot back at you. They won't be ass-hopping around on a hill with their ears flopping."

We had been trained as soldiers, but we were young and had never been in real combat. So, to us, death was still somewhat of a game, and popping a bunch of rabbits on the side of a hill seemed like fun. I would never have thought it at the time, but there would soon come a day when I would be thankful we had spared those lives, even if they were just rabbits. In the months ahead, we would each learn that there is enough death and destruction to go around in a war without killing for sport or pleasure.

8

Inspiration

Oxford, England
May 24, 1944

None of us knew much about Oxford when we stepped off the train, except that it had been designated an open city, which meant that it was supposedly exempt from getting bombed by the Germans. Silverman explained that for every open city in England, there was a counterpart in Germany. Each side would therefore locate the majority of its hospitals in these particular towns. Since we were setting up our

92nd bivouac area outside
Oxford, England, May 1944.

camp in a field at town's edge, I was glad to hear that Oxford wouldn't be getting attacked—and hoped that somebody had informed the Germans.

Besides having military hospitals, Oxford was also home to the famous university, which Averitt, Duthie, and I explored at every opportunity. The setting of the campus was quite peaceful and scholarly, with springtime blossoms in abundance, ivy-covered buildings standing along the edges of well-manicured quads, and old professors slowly walking about pondering important academic issues. The whole atmosphere made me want to pick up a book—any book—and read.

Creating our campsite took all of one afternoon. We each had half a pup tent and were randomly assigned to team up with someone from the 92nd. I was teamed with Chandler, who took up most of the space—and oxygen—in the tent. There was an HQ tent, a couple of shower tents, and the all-important mess tent. Hodges, who loved the idea of radio broadcasting, rigged up a PA system so that we could hear Armed Forces Radio throughout the bivouac area each morning while we ate breakfast. AFR played music and gave updates on the war effort, as well as general announcements and reminders of how much the folks back home were counting on us. It was a good morale booster, temporarily getting our minds off the uncertainties that lay ahead. It also helped us forget the fact that we were eating the dreaded powdered eggs.

General Patton Speaks to the Invasion Troops
Near Oxford, England
June 5, 1944

On the morning of June 5, we were awakened with orders to don our dress uniforms and be ready to move out for a special assembly. Word was soon spreading that General Patton was here to give us a speech. Shoes were therefore shined to a higher gloss and knots were tied with tighter precision than usual. Each man wanted to look his sharpest on this day.

As First Sergeant Thomas marched us from our campsite out beyond the limits of Oxford, we were joined by what seemed like thousands of soldiers from other units, coming from every direction and parts unknown. All were in full-dress uniform, counting cadences, standing tall, marching proudly.

We turned to the right off the macadam road, then marched up and over a grassy hill. There, before us, lay a small, peaceful, green valley, at one end of which was a wooden platform, decorated with red, white, and blue buntings and crowded with military brass and local dignitaries, the likes of which included countless colonels, majors, and captains along with the mayor of Oxford and an entourage of women from the Red Cross. A military band, standing on the left side of the stage, was playing marches, and an honor guard, supposedly handpicked by Patton himself, stood at attention nearby.

Soldiers were streaming into the area from all angles as hundreds of military police supervised the orderly assemblage of the troops. Sergeant Thomas had timed it so that we arrived just ahead of the convening multitudes and were therefore granted prime seating space on the ground just in front of the stage. Two flags, one American, the other that of the general—red with three white stars on it—stood feet apart at its center. Before them was a solitary microphone on a stand. A captain tapped it repeatedly, occasionally saying, "Testing, one, two, three," over the loudspeakers at different volumes.

The ground sloped slightly upward from the platform, stretching out until it rose into a hillside, creating a natural amphitheater. A sea of soldiers soon filled the entire area, some sitting high up on the hill, some even perched in trees so as to get a better look at the events of the day. The band was playing "The Stars and Stripes Forever." There was an electricity in the air of this beautiful, fresh English morning as we awaited the arrival of the general.

Finally a shiny black limousine, escorted by a convoy of jeeps, appeared on the road leading from town. A hush came over the troops, who now stood at attention. Those on the stage craned their necks in order to get a view of the entourage arriving behind them. The vehicles had barely come to a stop when General Patton emerged, his silver helmet gleaming in the midmorning sun, his knee-high boots shined to perfection, his ivory-handled pistol swinging from the holster at his side. He turned back to the car and pulled on a leash, attached to which was a short, humorless, white pit bull. The dog seemed determined not to dismount the vehicle, but a quick jerk from the general and the canine acquiesced, then led his esteemed owner up the steps onto the platform, where Patton sat in the seat of honor. General Simpson, General Haislip, and General Cook took their places next to Patton.

As the band finished playing, General Simpson approached the microphone. "Men," he said, "soon we will be fully engaged with the enemy." His words echoed across the landscape as each of us intently watched the most famous general still seated behind him. "You are well trained and prepared for the task at hand, and I have no doubt you will dispatch the enemy with true American courage and explosive power. Today you will have the honor of listening to the words of a great American soldier who will lead you in battle, one who has proven time and again that he can and will strike fear into our enemies while leading his own brave soldiers to victory—Lieutenant General George S. Patton Jr."

General Patton stood, handed the leash to one of the Red Cross women seated near him, then walked to the microphone. We were still standing at attention. The general looked out, seemingly at each man, inspecting from the platform, searching, trying to decide if he approved of those assembled before him. "Be seated," he commanded. In an instant, thousands of men were on their asses. *That's power*, I thought.

"Men," he began, his voice resonating throughout the valley with supreme confidence, "this stuff that some sources sling around about America wanting out of this war and not wanting to fight—it's a crock of bullshit. Americans love to fight traditionally. All real Americans love the sting and clash of battle. You are here today for three reasons and three reasons only. First, you are here because you want to defend your homes and your loved ones. Second, you are here for your own self-respect, because you wouldn't want to be anywhere else. Third, you are here because you are real men, and all real men like to fight.

"Men, when you here, every one of you, were kids, you all admired the champion athlete, the fastest runner, the toughest boxer, the big-league baseball players, and the best football players. Americans love a winner. Americans will not tolerate a loser. Americans despise cowards. Americans play to win all of the time. I wouldn't give a hoot in hell for a man who lost and laughed. That's why Americans have never lost nor will ever lose a war, for the very idea of losing is hateful to an American."

He shook his head slightly as he peered out over the silent audience. "You are not all going to die. Only two percent of you right here today will die in a major battle. But death must not be feared. Death, in time, comes to all men." He held his head high, as if in defiance of that very inevitability. "Yes," he continued, "every man is scared in his first battle. If he says he's not, then he's a liar. Some men are cowards, but they fight the same as the brave men, or they get the hell slammed out of them while watching others fight who are just as scared as they are. The real hero is the man who fights even though he is scared. Some men get over their fright in a minute under fire. For some it takes an hour. For some it takes days. But a real man will never let his fear of death overpower his honor, his sense of duty to his country, and his innate manhood.

"Battle is the most magnificent competition in which a human being can indulge. It brings out all that is best and removes all that is base. Americans pride themselves on being he-men, and they *are* he-men. Remember that the enemy is just as frightened as you are, and probably more so. They are not supermen.

"All through your Army careers, you men have bitched about what you call chickenshit drilling." A tentative chuckle went through the crowd. "That," he went on, "like everything else in the Army, has a definite purpose. And that purpose is alertness. Alertness must be bred into every soldier. I don't give a fuck for a man who's not always on his toes," he boomed as several of the Red Cross women onstage stared wide-eyed at each other, then down at the ground.

Patton looked briefly around in the direction of the women, seemingly to apologize for his outburst, but then thought better of it. He quickly turned his attention back to us. "You men are veterans, or you wouldn't be here. You're ready for what's to come. A man must be alert at all times if he expects to stay alive. If you're not alert, well, sometime or another some German son of an asshole bitch is going to sneak up behind you and beat you to death with a sockful of shit!"

We laughed out loud as the general nodded slowly. "Listen, there are four hundred neatly marked graves somewhere in Sicily," he said. "All because one man went to sleep on the job." There was no more laughter. "But they are German graves, because we caught the bastard asleep before they did!"

We jumped to our feet and cheered. Patton grabbed the microphone and yelled over our applause. "An army is a team! It lives,
sleeps, eats, and fights as a team! This individual heroic stuff is pure
horseshit! The bilious bastards who write that kind of stuff for the
Saturday Evening Post don't know any more about real fighting under
fire than they know about fucking!"

By this time we were going crazy, cheering, clapping, and laughing.
The general once again glanced around at the women onstage, this
time with a slight smile on his face, almost as if to make sure that his
comments had registered with them. For their part, they looked like a
bunch of nervous Nellies, almost frightened, almost indignant, but
mostly not knowing how to react.

Patton looked back at his men assembled in the field and motioned
for us to sit again. "We have the finest food, the finest equipment, the
best spirit, and the best men in the world!" He peered out over the
crowd, then thundered into the mike, "Why, by God, I actually pity
those poor sons of bitches we're going up against! By God, I do!"

Once again we all cheered and yelled in agreement. The general
seemed to be enjoying it as much as we were. "My men don't surrender,"
he continued. "I don't want to hear about any soldier under my command being captured unless he's been hit . . . unless he's been wounded
and dying and can't fight back. Hell, even if you are hit, you can still
fight back. And that's not just bullshit either. The kind of man I want in
my command is just like this brave lieutenant in Libya. He had a Luger
against his chest, but he jerked off his helmet, pushed the gun aside
with one hand, and then busted the hell out of the kraut with his helmet. Then he jumped on the gun and went out and killed another German before they knew what the hell was going on. And all of that time,
this soldier had a bullet through his lung. Now, there was a real man!

"Men, all of the real heroes are not storybook combat fighters
either. Remember that every single man in this Army plays a vital
role. Don't ever let up. Don't ever think that your job is unimportant.
Every man has a job to do, and he must do it. Every man is a vital link
in the great chain.

"Hell, what if every truck driver suddenly decided that he didn't
like the whine of those shells overhead, turned yellow, and then
jumped headlong into a ditch? The cowardly bastard could say, 'Hell,
they won't miss me, just one man among thousands.' But what if

every man thought that way? Where in the hell would we be now? What would our country, our loved ones, our homes, even the whole world, be like? No, goddamn it! Americans don't think like that. Every man does his job. Every man serves the whole. Every department, every unit, is important in the vast scheme of this war. The ordnance men are needed to supply the guns and machinery of war to keep us rolling. The quartermaster is needed to bring up food and clothes, because where we're going there isn't a hell of a lot to steal. Every last man on KP duty has a job to do, even the one who heats our water to keep us from getting the GI shits."

He looked around at the ladies and politely smiled before continuing. "Each man must not think only of himself but also of his buddy fighting beside him. We don't want yellow cowards in this army. They should be killed off like rats. If not, they'll go home after this war and breed more cowards. The brave men will breed more brave men. Kill off the goddamned cowards, and we'll have a nation of brave men."

The general lifted his gaze over the crowd as if his thoughts were far away. "Ah, you should've seen those trucks on the road to Tunisia. What a sight that was! Those drivers were magnificent. All day and all night, they rolled over those son-of-a-bitching roads, never stopping, never faltering from their course, with goddamned shells bursting all around them all the time. We got through on good old American guts. Many of those men drove for over forty consecutive hours. And these weren't even combat men, but they were soldiers with a job to do. They did it, and in one hell of a way they did it! They were part of a team. Without team effort, without them, the fight would have been lost. All of the links in the chain pulled together, and the chain became unbreakable.

"Now, don't forget—you men don't know that I'm here. No mention of that fact is to be made in any letters. The world is not supposed to know what the hell happened to me. I'm not supposed to be commanding this army. I'm not even supposed to be in England right now. Let the first bastards to find out be the goddamned Germans! Someday very soon I want to see them raise up on their piss-soaked hind legs and howl, 'Jesus Christ, it's the goddamned Third Army again and that son-of-a-fucking-bitch Patton!'"

We roared in approval as he continued. "We want to get the hell over there! The quicker we clean up this goddamned mess, the quicker we can take a little jaunt against the purple pissing Japs and clean out their nest, too—before the goddamned Marines get all the credit!"

I took my eyes from the stage and glanced around. Guys were cheering, clapping, yelling. We could tell we were about to become makers of history, and our leader was standing before us.

He waited for us to calm down a bit, then said, "Sure, we want to go home. We want this war over with. And the quickest way to get it over with is to go get the bastards who started it. The quicker they are whipped, the quicker we can go home. The shortest way home is through Berlin and Tokyo. And when we get to Berlin," he growled, gripping the ivory-handled pistol at his side, "I am personally going to shoot that paper-hanging son of a bitch Hitler just like I'd shoot a god-damned snake!"

We were hooting and applauding. "Now, listen," he continued, "when a man is lying in a shell hole, if he just stays there all day, a German will get to him eventually. The hell with that! The hell with taking it! I don't care what your instructors have told you—my men don't dig foxholes. I don't want them to. Foxholes only slow up an offensive. We keep moving. And don't give the enemy time to dig one either.

"We'll win this war, but we'll win it only by fighting and by showing the Germans that we've got more guts than they have—or ever will have. We're not going to just shoot the sons of bitches. We're going to rip out their living goddamned guts and use them to grease the treads of our tanks. We're going to murder those lousy Hun cocksuckers by the bushel fucking basket. War is a bloody, killing business. You've got to spill their blood, or they'll spill yours. Rip them up the belly. Shoot them in the guts.

"Let me assure you of this, men: When shells are hitting all around you, and you wipe the dirt off your face and realize that, instead of dirt, it's the blood and guts of what once was your best friend beside you, you'll know what to do!

"I want you to remember that no poor bastard ever won a war by dying for his country. He won it by making the other poor dumb son-of-a-bitching bastard die for *his* country. Remember that. And another thing: I don't want to get any messages saying, 'We are hold-ing our position.' We are not holding a goddamned thing! Let the Ger-mans do that! We are advancing constantly, and we are not interested in holding on to anything, except the enemy's balls. We are going to twist his balls and kick the living shit out of him all of the time. Our basic plan of operation is to advance and to keep on advancing regardless of whether we have to go over, under, or through the enemy. We are going to go through him like crap through a goose, like shit through a tinhorn!"

We were on our feet again. An electric charge had passed from the great general to all of us in the audience. Now it was surging through our beings, making us jump and yell a primordial scream. We wanted to run out at that very moment and defeat the enemy.

With a mere wave of his hand, Patton calmed us down. "Now, men, from time to time there will be some complaints that I am pushing my people too hard. Well, I don't give a good goddamn about such complaints. I believe in the old and sound rule that an ounce of sweat will save a gallon of blood. The harder we push, the more Germans we kill. And the more Germans we kill, the fewer of our men will be killed. Pushing means fewer casualties on our side. I want you all to remember that.

"And remember this part, too." He paused. "There is one great thing that you men will be able to say after this war is over and once you're home again. You'll be thankful that, twenty or thirty years from now, when you're sitting by the fireplace with your grandson on your knee and he asks you what you did in the great World War II, you won't have to cough, shift him to the other knee, and say, 'Well, your granddaddy shoveled shit in Louisiana.' No, sir, you can look him straight in the eye and say, 'Son, your granddaddy rode with the Great Third Army and a son of a goddamned bitch named Georgie Patton!"

And with that he turned and strode from the microphone back to the dignitaries onstage. We were going crazy, jumping with fists held high, slapping each other on the back, shaking hands, and cheering. I watched the general as he nonchalantly took the leash from the long-since-stunned Red Cross woman seated behind him, nodded to her politely, then led his dog across the stage, down the steps, and back to his waiting vehicle. The band struck up a rousing Sousa march, and all the men began clapping in time with it as Patton and his entourage disappeared into the English countryside.

D-Day
Oxford, England
June 6, 1944

The day began normally enough. Up early, exercise and a shower, then on to breakfast, where Chicago and Cardini were getting into their usual shoving match in line. Chandler was stuffing extra rolls into his pockets, Silverman was complaining about Tex, Tex was complaining about everything, and Sarge was trying to keep the peace and eat.

"Hey, Averitt," I said as I set my tray down on the table, "you and Duthie want to go into town and see a movie this afternoon?"

"Yeah, I'll go," answered Averitt.

"Sounds good to me," said Duthie, already seated across the table.

"Good," I said. "I'm in."

"Let's go over to the college first to see if we can find some girls to go with us," Averitt suggested.

"Hmmm . . . good idea," I said.

Just then we heard some commotion over the loudspeakers. As usual, Hodges had been playing Armed Forces Radio, but now he seemed to be turning it up louder. The announcer was introducing General Eisenhower.

"Citizens of Europe," began Eisenhower in that unmistakably rich baritone voice, "your day of liberation is at hand."

Everyone froze in place as we listened to the words echoing throughout the camp. The invasion had begun.

The next morning U.S. Army trucks began rolling past our camp, transporting the first wounded American soldiers from Normandy back to the hospitals in Oxford. After breakfast a few of us decided to go into town to see if we could be of some assistance. Sarge gave us some passes and told us not to get in the way.

We followed alongside the procession of trucks until we arrived at one of the many hospitals. Some of the doctors had set up a triage center on the grass outside the building, so we asked them if they needed us to help. "Just . . . just wait over there," one of them said, pointing to an open area. "We might need some help later. Not right now. Thanks." He looked busy, hurriedly checking the condition of wounded soldiers as they arrived.

We went near the street and began watching as truck after truck drove by. The tarps were off, and in the back of each vehicle, three or four soldiers lay crossways. Some were leaning on one elbow, smoking, looking like they were in shock. Others were unconscious and looked to be hurt pretty badly. Some had medics beside them, holding up bags of blood. Some were missing a leg or had large, bloodstained bandages around their abdomens or heads. We watched as the trucks slowly filed past us, but down the line—myself, Silverman, Duthie, Averitt, Chandler, Hodges, Cardini—none of us said a word. This was our first glimpse of the reality of war, and it was all too clear that it was only a preview of what we would soon be facing ourselves.

9

Countdown to History

Southampton, England
June 11, 1944

By the time we arrived in Southampton, it had become a staging ground unlike any in military history. Hundreds of thousands of men and untold tons of equipment and supplies had been assembling here for weeks. But now that the invasion, known as Operation Overlord, had begun, the activity and confusion had multiplied. Soldiers, sailors, and officers were everywhere, some busily working to load ships, some sitting, watching, waiting to leave, and some completely lost amid the bustling maze of streets near the water's edge.

At the docks dozens of huge ships were being loaded with men, vehicles, weaponry, and heavy equipment. Farther out in the harbor, hundreds of other boats of all sizes awaited their turns. Thousands of blimplike barrage balloons were hovering high over the channel, their vertical steel cables designed to keep enemy aircraft from strafing us. And above it all was the most ominous sight I had ever seen: The normally bright morning sky was dark ,with a massive fleet of bombers and fighter planes flying outbound from England to France.

We had been convoyed to one area of town and then marched to another, all the while trying to keep up with First Sergeant Thomas, who was intently studying a set of orders and a map as he led the way through the beehive of activity. We finally reached a large, crowded warehouse-type building, where Sarge counted us to make sure nobody had gotten lost.

"Okay," he said, "we're gonna be processed here, so stick together a few more minutes. Don't go wandering off with another unit. It'll only take a minute. I need to turn in the paperwork. Then we'll eat. According to my map"—he turned the map several directions as he studied it again, then pointed—"the mess hall is just next door. Stay right here."

And with that he disappeared into a sea of humanity in the dimly lit, musty building. Within minutes he had returned. "Chow time," he said, pointing for us to follow him. Then he held up one hand to stop us. "Look," he said, "after chow you're on your own; you're free to walk around. I don't care what you do. Don't get in anybody's way, don't get in any fights, and don't get lost. Team up in groups of three or four, and keep up with each other. Evening chow: same place we're going now. After that find a place to sleep. There's an area with cots over there and another area upstairs inside this building. We meet here—right here—in the morning. We move out at 0600 sharp. What time, Cardini?"

"At 0600, Sarge. I heard you."

"Sharp. Look alive, men."

A group of us spent the afternoon exploring Southampton. It was an impressive balance of logistical coordination and mass confusion. New units were arriving by the minute and moving out just as quickly. Fully loaded ships were being towed into the harbor to wait while others were moving toward the docks. We were given no information regarding what was happening at Normandy, whether we were winning or losing. All we knew was that we were next.

Silverman, of course, had some theories. He told us that we had originally been scheduled to be in the initial assault on the beach, but that General Haislip, the commander of the XV Corps of the Third Army, had gone to Eisenhower and told him something to the effect of, "Hell no, Ike. These damn boys will get slaughtered, and then we won't have any communications at all." According to Silverman, headquarters had agreed, and we were held back a week.

As improbable as such a story may have sounded, we knew that Silverman could have been correct, especially since his source was a sergeant in the 92nd named Mike Carapella. Sergeant Carapella was in charge of communications between all sectors and regularly hooked up calls between Eisenhower, Patton, Haislip, and the other generals while we were in England. He was a good man to know, especially if you wanted to find out what was going on.

After sunset the sky came alive with V-1 rockets launched by the Germans in France. The missiles missed the port entirely, flying overhead toward the northeast, following the general direction of the river toward the populated areas of town. A British antiaircraft crew stationed at the end of the docks fired repeatedly at the rockets but never actually hit any. When each rocket eventually ran out of fuel, the flame shooting from its tail would disappear and it would arc down and crash. Some of the American gunners were timing the difference between the visible flash and the sound of the explosion in order to calculate the distance away from us the bomb had dropped. According to them, most were falling in open fields north of town.

Watching the V-1s was a good show, but the morrow loomed as an important day for us, so we found our way back to the cot area for a final night's sleep before our official entry into the war. As we were getting ourselves squared away for the night, Chicago spotted a familiar face among the hundreds of men in the vicinity. "Hey, look, it's that guy from the boat who tried to grab Sacco's buster!"

He was only about eight cots away and clearly within earshot. He immediately turned and looked in our direction. "You're the queer from the boat," Silverman clarified.

"What are you doing here?" Averitt asked. "We thought they shipped your ass back home after that."

The boy shook his head and waved his hand in resignation, then laughed. "Hell, I ain't no queer! I was just doing that hoping that they'd send me home. I had tried everything else."

"Hell," Averitt responded, "if it were that easy to get sent home, we'd have been grabbing each other's privates all the way over here."

"Sounds like a pretty desperate measure to me," Spotted Bear said. "What if Sacco here had been a queer? Then you'da been in real trouble."

I instantaneously slapped Spotted Bear upside the head. "Look," I said to the guy from the boat, "if you really need to get friendly with someone . . . well, Bird Turd over there, he *is* a queer."

Bird Turd looked up from the *Stars and Stripes* he was reading. "Huh?"

"Nothing," I answered.

"Nobody said a thing," Spotted Bear reassured him.

"All of you can go to hell," Bird Turd mumbled before returning to his newspaper.

"Don't worry," the suspect guy said, putting on a woman's voice. "None of you guys are my type."

"That's good to know," Averitt laughed. "That's very good to know."

Just after breakfast the next morning, we boarded a ship called the *Philip F. Thomas*. We carried our backpacks, duffel bags, and rifles, of course, but all our heavy equipment, including the trucks, was winched aboard and stowed belowdecks. The process took most of the day.

They told us that there were hammocks below, but the weather was warm, so Averitt, Duthie, Chandler, Cardini, Silverman, Hodges, and I decided to stake out a place up on deck. Soldiers were streaming aboard, covering what seemed like every square inch of the ship. We eventually found a space on the bulkhead, just forward of the superstructure, where we put down our gear.

Late in the afternoon, when the ship was finally loaded to capacity, we were towed away from the docks into the Channel, where we dropped anchor for the night. Some of the guys were being loud and obnoxious, probably overcorrecting for their own fears and uncertainties. Others were sitting looking at the water, not saying a word. Others were smoking cigarette after cigarette after cigarette. Others were reading small Bibles or frantically writing letters. Some looked to be on the verge of tears. Others looked not to have a worry in the world.

Most of us were scared but determined, especially after the speech by General Patton. We knew we were in for a battle, but we were sold on the idea of making the other poor dumb bastards die for *their* country, as he had advised, so that we could come back home to tell our children and grandchildren all about it.

In the evening, after some gambling, drinking, and exploring of the ship, we went back to claim our spots for the night. We were sitting around talking when a porthole in the superstructure opened just behind Chandler. A black man stuck his head through the hole. "Hi there!" he said, smiling. He then disappeared and shut the porthole.

"What the hell was that?" Chandler asked, looking around.

"Some guy," I answered.

"Looked like Louis Armstrong to me," added Silverman. "But skinnier."

"What was he doing?"

"I don't know," I said. "Just saying hello I guess."

"Did he have a trumpet with him?" asked Averitt.

The porthole opened again, and the man, who did in fact resemble Satchmo, reappeared. "Hi there," he said again, with a toothy smile. "Hey, pssst," he began in a loud whisper as we instinctively leaned toward him. "You guys hungry?"

"Yeah," we answered in unison. Hungry? We were starving.

His head disappeared again, and he began handing us large, meaty roast beef, turkey, and pastrami sandwiches through the window. We thanked him, took the sandwiches, and feasted. He returned a few minutes later to give us bowls of ice cream. Providentially, we had bedded down just outside the officers' mess, and this fine gentleman was kind enough to serve us some of the leftovers.

Before long the German V-1 rockets reappeared in the sky on their nightly flights from France to Southampton. The gunners on board began firing at them but were missing. Finally they hit one as it flew directly overhead. The bomb exploded, and the concussion from the blast caused the ship to bob around in the water like a toy boat. More rockets screamed toward England as the night progressed. Now, buoyed by the one hit, the gunners were even more aggressive in their pursuit of the missiles. The whole affair was loud and somewhat exciting. Some of the guys watched the proceedings deep into the night. As for me, I put some toilet paper in my ears, shut my eyes, and, as the ship slowly rocked back and forth, kept thinking what it must be like back home. And I eventually drifted off to sleep.

Ed Duthie, Paul Averitt, and me in Normandy, France, June 1944.

PART III

Days of Battle

When the great day of battle comes, remember your training. And remember, above all, that speed and vigor of attack are the sure roads to success and that you must succeed. To retreat is as cowardly as it is fatal.

Battle is not a terrifying ordeal to be endured. It is a magnificent experience wherein all of the elements that have made man superior to the beasts are present. Courage, self-sacrifice, loyalty, help to others, and devotion to duty. As you go in, you will perhaps be a little short of breath, and your knees may tremble. This breathlessness, this tremor—they are not fear. It is simply the excitement which every athlete feels just before the whistle blows. No, you will not fear, for you will be borne up by proud instinct and inspired by magnificent hate.

—GENERAL GEORGE S. PATTON

10

Normandy

We awoke beneath skies once again darkened by Allied planes as the *Philip F. Thomas* steamed toward France alongside hundreds of other ships. Some boats seemed to be sailing so close together that a determined soldier could have jumped from one to another until he crossed the entire fleet. Gunships, battleships, transport ships, troopships, gigantic and small, fast and lumbering, all covered with soldiers, all making their way to the battle with purpose and pride. For a farm boy from Alabama, it was an impressive sight, seeing this great and historic armada in full stride.

Other than what we had read in the *Stars and Stripes* (which wasn't much), we were unsure of what had happened to the preceding waves of troops, and therefore unsure of what might be awaiting us. We were sure, however, that the time for jokes, false bravado, and loud talk had passed. Most of the guys now quietly measured the seconds by writing or praying as the ship plowed through the smooth gray waters of the English Channel, bringing us ever closer to the opposite shore and our date with destiny.

A priest carefully worked his way through the jumbled crowd of men, stopping here and there to say a prayer and give blessings for battle. He climbed up on the foremost part of the superstructure, sprinkled some holy water, and blessed us all. Even guys who weren't Catholic were doing their best to make the sign of the cross.

I pulled out the crucifix Grandpa Sacco had given me and held it in my hand. I said some prayers that God would watch over me and all the boys in our company and that we would have victory. I was saying the Hail Mary when Silverman pointed off the bow of the ship. "Look," he said. "There it is." There, rising into view from beyond the horizon, was the coast of France.

With Normandy within sight, the ships of the convoy began the long process of slowing to a stop. Some of the sailors were lowering heavy fishnet ropes over the sides of the boat when we heard a series of loud explosions. Three of the ships, drifting to a stop in the Channel, had hit mines. Water, smoke, and debris were sent high into the air as the vessels jerked and recoiled, then convulsed in the water. Huge waves rippling from the explosions tossed nearby boats back and forth, up and down. Two of the stricken ships were a good distance off port side, the other about half a mile starboard, so it was hard to assess the damage from where we were. Each was far enough away not to pose an immediate threat to us. Yet each was close enough to be a sobering reminder that this was no training mission.

Once the anchor was dropped and our ship was at a standstill, Captain English, the commander of Company A, gave us the order to start loading into the landing crafts, known as Higgins boats, bobbing in the water below. The rectangular boats were made of wood, and each could hold about three dozen soldiers, along with a three-man crew. While I was waiting my turn to go, a lieutenant from the 92nd named Shanahan put his hand on my shoulder. "Just want to say . . . nice to have met you, Sacco," he said with resignation in his voice.

"What the hell are you talking about?" I turned and asked.

"Lieutenant Shanahan here thinks we're all gonna get killed today," Spotted Bear pointed out.

"Just wanted to say that it was nice to have known you, that's all," Shanahan repeated. He looked toward the beach. "We probably won't make it."

"Like hell!" I said, tightening the straps on my pack. "I ain't planning on dying today or no day soon. I'm here to fight a damn war and then go home!"

"Yeah." Shanahan sighed, slowly shaking his head. "Whatever you say, Sacco."

"Damn, Lieutenant," piped an irritated Cardini. "Think you could be a little more optimistic? Jesus Christ!"

Climbing over the railing was no small task, considering the steel helmet, fifty-pound backpack, loaded carbine, and extra ammunition. The adrenaline surge of the moment, plus the not-so-gentle nudging of some guy's helmet on my butt, helped me clear the railing itself and swing down the other side into the netting. As I looked into the boats far below us—and the first reaction of all the guys was to stop and look down before proceeding—I saw some boys getting their feet tangled up in the webbing, flipping upside down, and hanging there. They flailed about, the weight of their gear pulling them backward off the ropes, as other heavy-laden soldiers tried to right them.

I was careful of every step, making sure my foot was secure on each rope before I put my full weight on it. It wasn't as easy as it seemed, because the whole apparatus was undulating with the cascade of men. The ropes would be one place when I looked down at them, then someplace else when I moved, making it very easy to step completely through the net and flip upside down.

To make a difficult situation worse, Averitt was right alongside me and was, as usual, talking. "You think the krauts can shoot us from here? Sons of bitches. I mean, hell, we can see the shore, so that means they can sure as hell see us. Whataya think? 'Cause they probably have high-powered rifles with high-powered scopes on them. Whataya think, Sacco?" I ignored him and moved on down the ropes.

Once I arrived in the landing craft, I went as far to the back as possible and lay down. I heard Duthie's voice. "Whatcha doin' back here, Sacco?"

"Layin' down."

"Yeah. Why? Seasick?"

"I figure fewer bullets can hit me this way."

In an instant Duthie, Averitt, Silverman, Cardini, and Chandler were lying beside me. Once full, our craft slowly pulled away from the ship, and we began the final journey to shore. I could see only whatever was directly above me—a series of dramatic clouds lingering from recent storms, under which was passing that relentless blanket of aircraft. Sarge's face did enter my field of vision at one point. He looked down at those of us horizontal, shook his head, and moved on.

The Higgins boats weren't exactly a smooth, comfortable, or quiet ride. Many of the boys—at least the ones who were standing—were getting seasick. The ride went on for some time. It didn't seem like the shore had been this far away when we scaled down the side of the ship. Then again, the longer it took to get there, the better.

Looking up into the passing sky, I had the feeling that I was completely lost and alone. I can never remember being so scared in my whole life, yet so quiet inside, so deep in thought. It's hard to describe what goes through one's mind at a time like this. I couldn't help but think of lots of things I wished I'd said and done before I left home. Of course, I thought about Mama and Papa and wondered if they knew where I was or what I was about to go through. And I thought about Grandpa Sacco and wondered if he was watching over me now. I pulled the silver cross from my pocket and again clutched it tightly to my chest.

"Ready to get off the damn boat!" yelled the sailor piloting the landing craft. As we scrambled to our feet, the boat came to an abrupt halt, as if we had hit something. The large iron front wall of the craft swung down, crashing into the water and sand below. Before us was a place the French called Calvados, a place we had code-named Omaha Beach. We were still about forty yards from the beach itself, but this seemed to be as close as we were going to get.

I put the cross back in my pocket, tightened the strap on my helmet, and then made sure all of my equipment was secured and ready. From the landing craft, we could see the immense amount of activity on the beachhead. Groups of soldiers were running in the direction of the smoky cliffs in the distance as trucks, jeeps, and tanks barreled along in the sand, some columns moving away from the water, others toward it. Sporadic gunfire and explosions pierced through the crashing of the waves. Some vehicles lay destroyed and smoldering on the beach. Others were partially submerged in the water, never having made it to shore. Landing craft of all sizes and shapes were unloading soldiers and equipment everywhere and then returning to the vast armada farther out in the Channel to pick up another load. And above it all were the planes.

"I said get the hell off!" demanded the sailor. Some of the guys up front ran down the gate and into the shallow water. They waded knee-deep for a few steps as the rest of us crowded toward the exit. Suddenly the ones in front completely disappeared beneath the waves.

"Oh, shit!" yelled the sailor. "Shit! Get the hell back on the boat!" We pulled back into the landing craft with record speed. Those submerged had to fight the weight of their gear in order to keep from drowning. A few struggled back into the shallows under their own power. The rest, including Spotted Bear, were pulled to safety by nearby comrades.

Once we were all safely back on the craft, the sailor shut the gate and put the engines in reverse. "Everybody here?" Sergeant Thomas asked as he counted.

"Shit!" the sailor repeated aloud. "Fucking sandbar! That's the third time that's happened!" He then steered the boat to the left, thrust forward, and pulled closer to the beach. The gate splashed down again. This time we were only a few feet from the shore.

"Go!" he yelled. "Get the hell out!"

Spotted Bear, near the front, was still gasping for air, trying to shake off water, heavy in his uniform. Most of the guys up there were the ones who had fallen into the water, so they were moving with a bit of caution.

"I said to get the hell off this fucking boat!" screamed the sailor. "I gotta go pick up more guys!"

Sergeant Thomas had worked his way to the front. Now he held up one arm and charged down the ramp, into the water, and onto the sand. Sarge was a former infantryman. He was used to stuff like this. "Let's go, men!" he said as we followed closely behind. Lieutenant Shanahan followed Sergeant Thomas like the rest of us.

I ran so fast that I don't even remember touching the water. The next thing I knew, I was on the beach, dodging the huge X-shaped tank traps, running to keep up with Sarge and the rest of the men.

"Keep low," Shanahan called out down the ranks. "There're still some snipers and pockets of resistance around here." An explosion shook stone from the cliffs to our left. He waved us to follow him up to the right near a group of American tanks. We gathered there and waited for our own trucks to land.

The beach was strewn with damaged vehicles and equipment, around which scurried hundreds of functioning vehicles and thousands of men on foot, all seeking higher ground as the tide rolled in. There were no bodies.

A convoy of familiar trucks and jeeps was approaching. "Hey, any of you fellers need a damn ride?" Tex drawled with a big smile as he drove up.

"All right, good, let's go!" Shanahan said as most of us jumped aboard the moving vehicles. The rest loaded onto the jeeps as we followed a path up the hill, past a couple of German strongholds destroyed by the infantry, then westward along the edge of the cliffs until we reached Ponte-du-Hoc, a fortress situated on a sheer rock high above the beach, seemingly impregnable, but bombed into submission by the American Navy and overrun by the Rangers.

Out in the Channel, as far as the eye could see, the great fleet of warships stood as the Higgins boats darted endlessly between them and the shore.

From there we turned inland, up into the highlands of Normandy, then northwest in the direction of Cherbourg, passing through the towns of Isigny-sur-Mer, Carentan, and Ste.-Mère-Eglise. Every town and village was destroyed, in shambles and burning. All along the way, we could hear German artillery firing at the front lines, only to be answered many times over by the Allied warships off the coast. Large craters and dead animals were scattered about everywhere, evidence that we were well within range of both sides.

As twilight approached, we came across a wooded area some distance outside Montebourg. The lead truck pulled to a stop as it neared the trees. We immediately dismounted from the vehicles, drew our weapons, and looked to Sarge for instructions. He moved up to the front and motioned for us to follow.

Three infantrymen, stationed near the top of the path, waved us forward. They looked tired and haggard. The youngest-appearing of the group was seated on a rock, smoking. His hand was bandaged, and he had blood and dirt caked on his face. "Signal Battalion," he said in recognition of our unit, taking the cigarette out of his mouth with his functioning hand only long enough to speak. "Good."

Another of the infantrymen looked down toward the end of our column before turning to Sergeant Thomas and the lieutenant. "Been taking a pounding here. Drove the Jerries back, but lots of losses."

We walked ahead into the zone of devastation. Everywhere I looked were strewn the bodies of American soldiers. Some were laid out peacefully, others in grotesque poses with gaping wounds and missing parts. Smoke from the battle still lingered in the air. We were stunned as we walked through the carnage, stepping over the bodies, peering through the haze, quietly proceeding. Amid the silence of the slaughter, it occurred to me that all of us had been trained to fight but none of us had been trained to die.

After a few minutes, I tried not to look at the entire panorama. I tried not to take notice of the scene for fear I would remember it too vividly. I stopped and became transfixed on a solitary dead soldier just inches from my foot. He was leaning against a rock, rifle in hand, shot through the heart, his head cocked back as if he were looking heavenward for one final lifesaving miracle. I slowly leaned over until I could see his dog tags. "McCarthy." He couldn't have been more than eighteen or nineteen.

"McCarthy," I whispered to myself. It struck me that young McCarthy's mother didn't yet know her son was dead. "Why did you have to die, McCarthy?" I felt a pain shoot through my heart as I wondered what my own mother was going through and how she would feel if she got news one day that I—

"Sacco! Get your ass away from that!" Sarge yelled. "Let's move the hell outta here, soldier!"

I took my gaze from McCarthy and looked around at Sergeant Thomas. "Get the hell over here right now!" he said, vigorously pointing toward the ground in front of him to emphasize each word.

I got up and moved in his direction, stepping over a couple of bodies along the way. As I approached, I pointed back to McCarthy.

"Put it outta your mind, Sacco," he instructed.

"But—"

"Put it outta your mind." He looked around at the other guys, many of whom were standing still, staring at the scene. "Okay, let's move out," he said quietly.

We camped a mile or two away in an area dominated by bushes, trees, and tall, canelike grass, about four or five feet high, all of which made it easy for us to blend into the landscape and hide. Then again, it also made hiding easy for the enemy.

Chandler and I were assigned guard duty that first night. From our position we could see great flashes of light and hear earth-trembling explosions to the left, then to the right, then in front, then behind. A series of strange-sounding planes flew overhead. Then it would get quiet. A burst of machine-gun fire would echo from a distant hill. Then the flashes of light would start again.

"Chandler?" I whispered.

"Huh?"

"What's the code word?"

"What?"

"What the hell's the code word?"

"Hell, I don't know."

"Captain gave us a code word, told us to remember it."

"Yeah, that's right," he said. "I remember."

"Well," I pressed, "what is it?"

"Hell, I don't know. I remember he told us not to forget it, but I don't remember what it was."

"Well, that's just great, Chandler," I said. "That's just great. Shit."

"So what is it?" he asked.

"Hell, I don't know," I answered. "I forgot. I figured *you'd* remember. I got a lot on my mind. Shit!"

We quietly peered out into the darkness a little while longer. "Look," I finally said, "if anything moves, let's just shoot it. To hell with the code word."

"I was planning on doing that anyway," Chandler responded.

Every rustling of the wind, every fluttering of a leaf pulled our attention in that direction. We were ready to shoot and ask questions later. About two hours into our watch, we heard the unmistakable sounds of something moving through the trees. Our senses became alert, gathering information, preparing for any eventuality. Though the trespassers were trying to be stealthy, the undergrowth crackled ever so slightly beneath their feet, and the tall grass gently swayed as they moved about. Something was definitely out there prowling through the night, slowly approaching. My heart was pounding as I lifted my carbine.

"Halt!" I called out. Whatever it was stopped momentarily, then continued moving. I looked over at Chandler, who had his rifle trained in the direction of the noises. "Halt!" I repeated as the adrenaline surged. "Stop or I will shoot your ass right now!"

"Bicycle." We heard a voice with a decidedly New York accent come from the trees.

"What the . . . ?" Chandler whispered.

"Bicycle," the voice called again.

"Silverman?" I said, looking over my carbine.

"Sacco?" Silverman and Chicago emerged through the grass, their weapons pointed at us. "I say 'bicycle', and you say—"

"Wagon." I finally remembered the code words. "What the hell are you doing out there? We could've shot your asses!"

"Coming to take over watch," Chicago said. "Whataya think? We were just out for a stroll?"

"Hey, Sacco," Silverman began, "you're supposed to say the code word, not 'halt.' That's a damn Nazi word. You're supposed to say the code word, and if we don't respond, then you—"

"Hey, Silverman," Chandler interrupted. "Shut up."

Heading back to get some sleep, we walked through several rows of pup tents. As we passed each, we could hear rifle bolts inside being snapped back, readying the weapons for fire. That was a bit unnerving. Suddenly we heard a great deal of noise back near the lookout post. A volley of shots rang out. Then quiet.

Captain English, Sergeant Thomas, and some of the guys came running in our direction. We turned and ran with them until we reached Silverman and Chicago, who still had their weapons pointed into the dark of night. We dropped to the ground and waited for the next sound. Besides the pounding of our hearts, all we could hear was a breeze swaying through the trees.

After a few seconds, Sarge motioned for Cardini to follow him. Cardini motioned for Chicago to go instead. "Cardini," Sarge angrily whispered, "get your ass up here right now. Follow me!"

The two crawled off into the high grass. They returned several minutes later. "Congratulations, Silverman," Cardini said. "You just killed a cow."

"Probably didn't know the code word, huh?" Chandler asked.

"You two, back to your post," Captain English instructed. "The rest of you, get some shut-eye."

"Go to hell," Silverman said under his breath.

"What was that, Private?" asked the captain.

"Talking to Chandler, sir," Silverman answered.

"Back to your post," English repeated. "And stop shooting the livestock, Silverman."

We moved out the next morning, up closer to Valognes, where we began setting up communications for the infantry. Now that we were surrounded by war, our training back in the States seemed simple and innocent. Technically we knew what to do, but the realities of actual combat—hearing the constant gunfire, feeling the earth shake as artillery shells exploded around us, seeing soldiers and civilians lying dead in smoldering ruins, smelling death and destruction with every shifting of the wind—all these conspired to have a chilling effect. Within twenty-four hours of our arrival in Normandy, we had witnessed devastation beyond anything we'd ever imagined. It was a shock to the system and not something to which a person of reasonable sanity can ever adjust.

Back in England most of the guys had composed letters, which we carried on us, that would be sent back to our parents if we didn't make it. After a couple of days in Normandy, I pulled out mine to make some revisions. I wrote "Dear Mama and Papa" atop a sheet of paper, stared at it for about half an hour, then crumpled it up, grabbed another sheet, and started over. I repeated the process several times.

Finally, Lieutenant Shanahan, who was sitting nearby, spoke up. "Sacco, you gonna write something or just waste all the paper?"

"Just one more sheet," I said, determined to get it right. I had the original letter before me, but after what I had seen so far, I felt there was more I should say. Death seemed more real to me now, more sudden and more permanent. I sat staring at the blank sheet.

"Hell, what do you care?" Shanahan ribbed. "You're gonna live through all this anyway, remember? You can just tell 'em all about it when you get back home."

"Yeah, that's right," I said. I folded the paper, got up, and walked away. I reached into my pocket and clasped the crucifix. *I hope you're right, Shanahan,* I thought. *I hope to hell you're right.*

As the battle for the city of Cherbourg heated up, so did the German attacks on our positions. The city, a major port on the English Channel, was of obvious strategic importance, and the Germans seemed willing to fight to the last man to keep it under their control.

Our infantry was taking some heavy losses, but the Nazis were taking more, being blasted out of each town, each village, and each encampment along the way to Cherbourg. Some towns were falling to us, then back to the Germans, then back to us. At least those were the rumors. And there were lots of rumors.

The only thing I knew for sure was that we were catching some hell as we worked to set up telephone lines. German planes would swoop down, drop bombs, and strafe as we ran for cover in every direction. We were constantly on the move, advancing and retreating with the infantry, making sure there were always lines of communication between the front and headquarters. Our job necessarily put us in harm's way on a continuous basis. Often we were actually in front of the front lines, high in trees, stringing wire while bombs blasted around us. At night the terror continued, with enemy artillery raining down on our position, lighting up the sky while digging huge craters in the earth. The explosions, the rumbling, the threat, and the fear were incessant.

The pressure got to be too much for some of the guys. One boy, a bigmouth from Philadelphia named Langford, couldn't take it and shot himself in the foot, hoping to get sent home. But his ploy backfired. Instead of a trip Stateside, he was given a court-martial and then reassigned to the infantry, bad foot and all.

Cherbourg finally fell toward the end of June. Though the Germans had promised to fight to the bitter end, the advancing American infantry and Allied air and naval bombardment had been too much. The Germans hightailed it, and our infantry took control of the city and the port, giving us a major victory at this early stage of the war.

That was the good news. The bad news was that the enemy tried to destroy as much of the city as they could before they left, sabotaging supply depots, ammunition warehouses, and fuel lines. As they fled, they planted land mines on the roads leading into town. Our infantry had captured several hundred German prisoners in Cherbourg, but many more had gotten away, meaning that there were now thousands of pissed-off Nazis hiding throughout the French countryside, waiting for a chance to kill us.

We could see thick black smoke rising from the city, almost obscuring it completely, as we approached. American planes pierced the dark billows and delivered more bombs to nearby German fortifications. The Germans were responding with artillery and antiaircraft fire. Occasionally we would hear a missile buzz overhead, then a massive explosion in the distance.

"My friends, that's the Navy airmailing a major ass kicking to somebody," Cardini said.

Tex, driving the truck but able to hear the conversation in the back, replied, "Uh-huh, but I hope they ain't sending it to us."

"It happens," Silverman said.

"What the hell are you talking about?" Chicago asked. "You saying the Navy decides to kick the shit out of us when they run outta krauts to bomb?"

"Sometimes they miss," Silverman continued. "Kill our own people. Happens all the time in combat."

"Hey, thanks for your uplifting observations, Silverman," said Cardini. "Now, shut the hell up."

Silverman began, "I'm just saying that sometimes the—"

He was interrupted by an explosion ahead of us that sent the lead jeep sideways and down a small embankment. The convoy jerked to a sudden halt. Medics ran up to the stricken vehicle, now engulfed in flames. There wasn't much they could do. Three men, including Lieutenant Shanahan, were dead.

We stayed in Cherbourg for only a couple of days—long enough to set up some phone lines for the occupation forces—before moving south to Barneville. As we passed through villages still smoking from battle, people emerged from the rubble to cheer, toss us flowers, wave flags, and chant *"Vive la France!"* Some were handing out small glasses of calvados, a potent local drink, of which we gladly partook. It was good to help these people celebrate. Hell, they'd been through a lot.

Some of the guys found a bar still standing in one town along the way and thought it would be the appropriate place to both continue the celebrations and relieve some tension. By the time the 92nd arrived, though, there were no bottles left on the shelves behind the bartender. Instead we found passed-out soldiers, drunk beyond recovery and now oblivious to the raucousness surrounding them.

"Kinda early in the day to be pickled, ain't it?" observed Averitt. It was well before noon.

I felt a tap on my shoulder. It was Sarge. "Soldiers, I don't know if you heard the news, but the war's not over yet. So let's move out."

"Just one drink, Sarge," pleaded Spotted Bear.

"We just got here," said Averitt.

"Move out," Sarge said again. "We got a war to fight. Come on. I got a bottle of bourbon for the road."

We were walking through the countryside behind the truck, wondering when Sarge was going to break out that bottle, when we heard an odd sound in the sky above. A strange, black, fast-moving aircraft was shooting across the horizon, heading west.

"Look at that," Averitt said to those around him, as if they might not have already spotted the plane.

"How the hell is it flying?" I asked.

Chandler couldn't figure it out either. "It ain't got no propellers!"

"What a bunch of ignorant, stupid, backwoods bumpkins," Silverman derided. "Of course there's no propellers. Geesh. Sarge," he called up the ranks, "do I have to be in the unit with these morons?"

"Shut up, Silverman," warned Chicago.

"*Shut up, Silverman*," Silverman mocked. "That's the best you can come up with?"

Chandler was getting fed up. "Okay, you little kike, why ain't it got no propellers?"

Silverman sighed and rolled his eyes patronizingly. "Because it's not a plane, you bunch of rubes. It's a German missile. It's a bomb. It's just got wings to help it stay level while it flies."

"Hey, Silverman," Sarge piped up, "it *is* a plane. It's called a 'jet' plane. It's got jet engines instead of propellers, so it can fly faster, higher, longer. Something the krauts came up with."

Chandler turned and, walking backward, pointed at Silverman with both hands. "Moron!" he laughed. Before he could turn back around, he tripped and almost busted his ass.

Barneville, France
July 2, 1944

We bivouacked later that evening in a wooded area outside the town of Barneville, near the west coast of France, where we reorganized with other companies and then waited for further orders. Chandler and I set up our tent just beside a large tree near the middle of the campsite. The location was technically no safer than any other, but the buffer of the other men and our proximity to the tree made us feel better.

All around us we could hear American artillery blasting away at the enemy and, of course, the enemy blasting away at us. The sounds seem to be magnified at night, perhaps because our sense of hearing was heightened by the absolute darkness. It was easy to feel vulnerable in the pitch black, lying on the ground, wound up in a sleeping bag, noises creaking outside the tent, bombs shaking the earth beneath us. On top of it all, I was in the tent with Chandler, who still had the potential of going berserk at any moment.

Each evening, just after darkness overtook the camp, we would hear the distinctive humming of German bombers, which we called "Bed-Check Charlie," flying over, searching for our positions. The discipline of total blackout and the ability to perform simple tasks without light was embraced and appreciated. This meant no headlights, no flashlights, no smoking, not even striking a match. One slip could give away our position and cost the lives of hundreds of men.

We were in bivouac for several weeks, during which time there were no duties other than eating and, occasionally, exploring. Of course, exploring wasn't the safest thing to do since there was this little matter of a world war raging around us. Averitt, Duthie, Hodges, and I ventured out a few times, finding some destroyed German tanks and other vehicles. That was fun. But the best thing we discovered was an American Army unit in charge of baking bread for the soldiers. We carried hot, fresh loaves back to our guys and were instant heroes.

There wasn't much to do during our stay in the area, but it certainly wasn't a vacation. The jitters of war were still fresh and emphatically heightened by the enemy at every opportunity. There were several bombing raids that scared the hell out of us, digging up large chunks of the camp, destroying vehicles, and sending shrapnel hundreds of yards in every direction. One particularly ominous strafing shot up our pup tents and plugged dozens of rounds into the tree beside which Chandler and I had camped. Fortunately, the attack took place at midmorning, and all the guys were either out of the area or able to run for cover.

Other encounters with the enemy were more deadly. A couple of jeeps from the 92nd hit land mines, killing or severely injuring their occupants. Then, at the end of July, three men from our company—Don Castillo, Leo Rimini, and Charlie Wright—were ambushed and killed while on a patrol. The news hit hard, because we knew these guys, and they were good men. For the rest of us, it was yet another sobering reminder that this was no game and that our time could come at any minute.

11

Breaking Through

Barneville, France
August 1, 1944

The sun was just beginning to illuminate the sky, and the canopy of planes that had accompanied us across the English Channel was once again aloft, this time reflecting the day's first golden rays and in a formation so vast that it stretched out far into the horizon. As the morning grew brighter, the sight grew more and more spectacular. Sarge told us there were probably three thousand or more bombers in the sky, paving the way for our breakthrough of enemy lines. Though they were dropping their payloads on targets miles away, we could hear the massive explosions and feel the rumblings underfoot.

Captain English had come around the night before with orders that we were to move out at first light. Now, with trucks packed and routes mapped, we began rolling into the French countryside, south from Barneville in the direction of Saint-Lô, where General Patton was already in the process of administering a massive ass kicking to the Germans. Our mission was to catch up with the infantry at the front lines and set up their communications.

We walked ahead of the trucks—which carried not only our work equipment but also our duffel bags and backpacks—holding our rifles at the ready, constantly scanning our surroundings for signs of danger. All around us were the sights and sounds of devastation: villages and farmhouses reduced to rubble—their occupants still somehow able to smile and cheer and wave flags to welcome us— fields pockmarked with burning vehicles and littered with dead and decaying animals, trees broken and smoldering, bridges destroyed.

And, echoing from every direction, the constant thunder of bombs.

We were walking along a road about a mile outside the demolished town of Périers, struggling to open our K rations for lunch, when we heard American fighter planes coming up over the hills to our right. They swooped down toward us, then began strafing and bombing something on the other side of a small hill just beyond our line of sight. We hit the dirt and took cover. The planes circled for another pass as machine-gun fire went up from the unseen valley. This time, the Americans unloaded an even heavier firestorm on their targets. And the valley grew quiet. They circled once or twice more, then disappeared beyond the hills.

Captain English sent a scout team up ahead to check out the situation. After a few minutes, they returned and gave us the all clear to proceed.

As we rounded the bend, we saw dozens of German trucks and jeeps destroyed and burning. Thick, black, acrid smoke billowed high into the air. Hundreds of German soldiers lay dead amid the wreckage.

"Oh, shit!" each guy said individually as the scene came into view.

"Oh, shit" was right. These Germans had been lying in wait, probably tipped off by a patrol that had spotted us on the road. They were prepared to ambush us as soon as we rounded the bend. But an American lookout somewhere in the hills had apparently seen the drama unfolding and had alerted the Army Air Corps to come and take them out before we became their prey.

As we walked through the carnage, I took notice of a jeep destroyed in the raid. One German soldier was dead in the back, his right hand still on the machine gun mounted there, his body twisted down into the seat. Two others had been blown through the windshield and were now partially draped over the hood. They, like most of the vehicle, were aflame. In my same field of vision, through the ripples of intense heat rising from the fire, I could make out Chandler, Spotted Bear, and Chicago on the other side of the jeep, staring at the spectacle, pulling food from their K rations, and eating.

We moved south through the countryside, ever mindful of dangers apparent and perils hidden. We were no longer exclusively following the roads but were going through fields, across ravines, up hills, over embankments—wherever and whatever it took to reach our objective. Captain English was constantly checking his map, looking at his compass, searching for landmarks, and then moving us forward. A small patrol was sent up ahead to scout the terrain.

Most problematic were the hedgerows lining many of the roads and dividing the fields. These weren't simply rows of hedges as the name might imply but were an intricately tangled weave of bushes and trees, dense and impenetrable, six feet high and four feet thick, into which were burrowed tunnels and fortifications. The Germans would hide like rats inside these shrubbery caves, then spring to attack their victims.

Weapons of destruction didn't seem to have any effect on the hedgerows. Artillery, mortars, and tanks would pound them, while the infantry sprayed them with lead, only to have the unscathed enemy hop up from within and open fire again. Using bayonets, infantrymen resorted to hand-to-hand—or in this case hand-to-plant—combat, pushing the enemy back hedgerow to hedgerow through Normandy.

Finally some guys in an armored division took some heavy pieces of steel and welded them onto the front of a Sherman tank. The tank then blasted its way through the hedgerow, opening a swath and sending Nazis running like roaches. We were told that the vegetation had, for the most part, been cleared of the enemy by the time we arrived. But most of us were still suspicious, keeping a cautious eye out whenever we neared them.

Sarge had us install our bayonets, so a few of the guys naturally insisted on poking the hedgerows as we passed. Bird Turd got his arm stuck inside one of them. The other guys were laughing as they walked by. "Come on!" he pleaded. "Cardini, Averitt, Sacco, hey, Duthie—somebody help me get my damn hand outta this."

Sarge was not pleased. "What the hell's your problem, soldier?" he asked, extricating Bird Turd's rifle from the bushes.

"I got my arm stuck in here, Sarge. Son-of-a-bitching piece of shit! Why the hell do these frogs have to put this shit all over their goddamn country?"

"Did that plant do something to you, Wilkins?" Sarge inquired as passing soldiers made an assortment of rather crude remarks.

"It's got my damn hand, Sarge!" Bird Turd replied.

Sarge, still holding the freed weapon, surveyed the rest of us moving down the road. "Young man," he said, turning back to Bird Turd, "you're a disgrace to the United States Army, that's what. Look at the rest of these men going along without a problem. And you—you gotta boondoggle and play around!"

"Sarge, Sarge," Bird Turd whimpered until he got Sarge's attention again.

"What the hell do you need now, Wilkins?"

"Sarge, can you help me get my arm outta this goddamn bush?"

Sarge turned and walked away. Somebody must have eventually cut Bird Turd free, because he and his injured arm were in camp with us that night.

We camped near the town of Fougères but only slept about three or four hours before moving out again. The infantry had opened a corridor to Laval, and our services were needed at the front. The different companies of the 92nd Signal Battalion had different duties: Company A (our company) and Company B were at the front lines, often between the infantry and the enemy, stringing wire back to Company C, which tested and repaired the lines before relaying everything back to HQ, which operated the main switchboards connecting the top brass to the action. We would rotate with Company B—one company up to the front, the other back a mile or so to get some rest.

The Germans were retreating, but they had this nasty habit of scattering into the countryside, regrouping, and then mounting a series of counterattacks on our positions. What would be our territory one afternoon would be the enemy's that night, then ours by daybreak the next morning. All the while there was hell to pay on both sides.

The fortunes of battle had a special significance to the boys of the 92nd in that we were stringing telephone lines in areas that would occasionally change hands while we were still on poles or in trees. This meant we had to get on the ground, load the trucks, and haul ass to catch up with the infantry. The Germans would come through and either confiscate the wire for their own purposes or cut it so that we couldn't use it in the future. That pissed us off.

Oftentimes, from our vantage points aloft, we could see American and German forces on each side of us maneuvering to prepare for battle. After a while the lines between safety and hazard became somewhat blurred, and we just went about our business. Infantrymen

Jim Hodges at a battlefield
switchboard in France, 1944.

regularly scolded us for being exposed to enemy fire and for giving
away their positions.

"Hey, buddy," we'd call down, "we got a job to do up here, so get off
our asses. Our job is to string line. Your job is to shoot krauts."

Averitt, Duthie, Silverman, Cardini, Sam Martin, and I were working in some trees one afternoon in an area outside Laval when a jeep
screeched to a halt below. I noticed Tex and Chandler on the ground
snap to attention and salute. An officer leaped from the vehicle and
looked up in our direction. It was General Patton. He walked briskly
to the base of Silverman's tree.

"Soldier!" he yelled up.

"Uh, yes, sir," Silverman answered, trying to salute but dropping a
pair of pliers instead. The tool hit several branches, causing the general to step out of the way as it fell to earth. "Yes, sir, General," Silverman nervously repeated.

"Soldier, what the *hell* are you doing up there?"

Silverman was hesitant. "Stringing phone line, sir."

"Don't you know you're a target up there in a tree like a damn monkey?"

Silverman looked around at the rest of us. We were absolutely still, not wanting to shake any leaves or have the general turn his attention to us. "Yes, sir, General, but—"

"But what?" Patton had a way of asking a question that made a soldier want to run.

Silverman took a deep breath. "But, sir, the infantry needs the phone lines, sir."

"Can't you see those sons-of-bitching Germans over there?"

"Yes, sir, General Patton." Silverman looked into the distance in the direction of the enemy. "I can see them real clear from here, sir."

"Well, then?" Patton asked. "Don't they make you nervous being up there, son?"

"No, sir," Silverman answered. "They don't. But *you* do."

"Ha!" General Patton laughed as he walked briskly back to his jeep. "Hot damn, Signal Battalion!" he roared, slapping his driver on the back. "Got the bravest, craziest sons of bitches in the whole goddamned Army!"

Once we hooked up with the Third Army, we saw Patton almost daily as he came up to the front to check on his troops. He would stand holding on to the windshield as his driver sped over the rough terrain, dodging obstacles to get him where he needed to go. Occasionally he would stop and ask how we were doing or just talk to us, telling short stories from his war experiences or giving personalized instructions on how to do something relating to combat. He was a serious character, and he asked a lot from his men, but we could tell that he had a genuine affection for us at the front lines, and we appreciated it. His concern and willingness to put himself in harm's way made it easier for us to accept the ever-increasing hardships of war.

Moving forward and working nonstop afforded us precious little time for such luxuries as eating or sleeping. The Army gave us K rations, not-so-tasty meals each consisting of a can of unidentifiable meat and a can of unchewable bread, which we often had to devour on the move. We slept on the ground, sometimes in tents but more often than not under the stars and flying artillery shells. I decided not to use my sleeping bag because I wanted to be able to run in case of imminent danger and felt a bit too vulnerable tied up in a sack on the ground.

Whenever possible, we sought shelter in barns or abandoned houses or even under the truck—a practice not encouraged by Captain English—to catch a blink or two of shut-eye. But even on nights when we had five or six hours before we were to move out, it was difficult to sleep. The constant thunder of bombs, occasionally rattling the timbers of whatever housed us, didn't create the optimal environment for relaxation.

And just when it seemed like things couldn't get any worse, it would rain. If the roads were difficult to traverse while dry, they were downright miserable when mixed with water. Thick, deep mud, churned and mashed by hundreds of trucks, made driving next to impossible and walking not much easier. Mud would stay caked on us for days, since there was rarely an opportunity to wash.

Supplies were running short, especially in light of the fact that the Germans continued the practice of watching us string wires during the day, then stealing them at night to alleviate their own shortages. Corporal Longley, who was Sarge's right-hand man, had the idea of swiping some wire from the top of one of the electric trains near the city of Le Mans. He climbed up on the train, took out some tools, looked down at us, and said, "Hell, we hit the jackpot, boys. There's lots of wire up here."

As he touched the wire, his body was sent into convulsions. He gripped the line and shook as we watched helplessly from below. Finally he fell onto the tracks.

A few of us ran to help, but when we tried to lift him from the ground, he came apart in our hands. Sarge called for a medic, but it was too late. He took one of Longley's dog tags and the letter from his inside pocket. He said a short prayer, then got up to walk away. "Get Captain English on the phone," he said without expression. "Let's move out."

We moved farther north in the direction of Argentan, a town on the southern edge of what was known as the Falaise Gap. The gap was occupied by German Panther units and constituted a major headache for Generals Eisenhower, Bradley, Montgomery, and Patton. The first three wanted to talk extensively about strategy before proceeding.

Patton, not one for politics or prolonged discussions when action would suffice, was determined to simply storm in, destroy the Panthers, and close the gap.

While Patton tried to convince Ike to let him do his job, we set up camp in the woods and awaited our orders. A quick bath in a small creek felt like a great luxury, as did eating a hot meal prepared in a real mess tent and then gathering a bed of pine needles under an Army blanket for our first good night's sleep in weeks.

Even so, not all was well in paradise. Many of the guys had developed nervous habits because of the pressures of war. Every now and then, right in the middle of a conversation, Chicago would start blinking repeatedly, as if he were trying either to protect himself from some intense, flashing light or to count to a hundred using his eyelids. The first time I saw him do it, I thought he was trying to send me a message in Morse code. Then, after a while, he would just calm down and look around like nothing was wrong. When Cardini got nervous, he would chew on the inside of his lip, sometimes so hard

Hodges takes a quick bath in a stream in France, 1944.

that it would bleed. Chandler began using water from his canteen to brush his teeth five or six times a day; not that that was a bad thing in itself, but it was a little odd for a man who took a bath once every three weeks—even in peacetime. Spotted Bear would sporadically check the inside of his helmet and poke around as if some small animal might be nesting in it. Averitt, when he wasn't jabbering, was tapping his leg . . . and tapping his leg . . . and tapping his leg. As for me, I kept a firm grip on my grandfather's crucifix and Mr. Sciatta's nuts, not that I was necessarily superstitious regarding the nuts, but I thought it better to be safe than sorry. There was, among the rest of the men, the expected assortment of twitches, tics, and furrowed brows. And all of us were smoking cigarettes like there was no tomorrow.

Chicago was still antagonizing Silverman, repeatedly calling him a kike and telling him that all this was the fault of "your people." Silverman finally took a swing at him, and they got into a brawl. Chandler stepped in, stopped the fight, and bluntly threatened to kill both of them if it happened again.

In addition to the many internal dynamics at work within our ranks, we were also greeted by the occasional unexpected guest. Johnny O'Brien, a guy in our company from California, looked

Here I am taking a break in France, 1944.

around one morning to find a young German soldier sleeping next to him. O'Brien woke up the intruder by pointing a carbine between his eyes and saying, "Hey, kraut!"

Using broken English and hand signals, the Nazi did his best to explain that he wanted to surrender. O'Brien, unsure of what was happening, started calling for Sarge. When we heard the words, "There's a damn kraut in here," several of us ran in his direction. By the time we arrived, O'Brien was crouched down outside the tent, his rifle still pointed at the German, who was crawling out on his hands and knees, doing his best to hold a white piece of cloth in front of him. Every gun in the vicinity was locked on the frightened German. Sarge helped him to his feet, and a couple of the guys confiscated his weapons and escorted him away.

Later that same afternoon, five Russian soldiers wandered into our camp saying they wanted to defect to America. Because they were technically our allies, they were taken to Captain English, who interviewed them and then gave them some American uniforms to wear until they could be transported back to headquarters. When we saw them walking around in our garb, we raised more than a little bit of hell. We knew they were supposedly on our side, but they weren't American soldiers and we didn't appreciate their wearing the uniform we had earned. We complained to Sarge, and he complained up the chain of command. Captain English then ordered that the Russians' uniforms be dyed black so that nobody could mistake one of them for one of us.

Despite all the extracurricular activities back at camp, we still managed to set up some phone lines for the infantry in preparation for battle. Finally, about the middle of August, the armored divisions began punching holes in the German defenses around the Falaise Gap. Patton had gotten his way—now he was blitzing the gap from all angles, raining down heavy, powerful, and deadly force on the enemy.

We followed the infantry until we stood amid the smoking ruins of what had been the city of Falaise. Destroyed enemy vehicles, some overturned and most still on fire, clogged the avenues of retreat. Some American infantrymen were transporting dead Germans in trucks and then dumping them in stacks on the side of the road, while others herded hundreds of Nazi prisoners through what was left of the town.

Duthie, Averitt, Spotted Bear, and I were watching the events unfold when we noticed an officer's jeep slowly making its way down the street. It was General Patton, standing, surveying the scene, intently studying each detail of the destruction as he passed. He looked in our direction, and we saluted. He saluted back, then slightly smiled and nodded in approval. The man was a warrior. He was here to kick ass, and we loved him for it.

12

East Through France

August 14, 1944

We didn't really have much information about the overall war effort other than what we heard from the infantry or other units whenever they were passing through. Most of what we heard—and repeated—were rumors, nonverifiable, unsubstantiated, not necessarily true, but always interesting rumors. And, according to the rumors, we were winning.

We were heading east, pushing the enemy back with even greater vigor. But it wasn't as if the Germans were giving up without a fight. In fact, they were attacking with ferocious determination and deadly force, be it via assaults from ground troops holed up in towns and villages, bombing raids from the Luftwaffe, or massive artillery shelling from deep inside their territory. Every town was a new battle. Every building, every barn, every tree, every bush was a potential hiding place for the enemy—and therefore a threat. And though battles were still raging and our guys were still dying, we had the feeling that we were conquering inch by inch, block by block, as we fought our way east through France.

We came to a wooded area where Sarge said we were to string some wire in anticipation of the arrival of Company B, who would splice our cable onto theirs and then route it back to HQ. The shelling was relatively distant on this day—close enough to hear but far enough away so as not to pose an immediate threat. It was around noon, and we had a few minutes before we needed to get to work, so we pulled the K rations from our packs, opened them using the small but trusty Army-issue P38 can opener that each of us carried in his pocket, and began the process of choking down another meal.

"Look at these tracks," said Spotted Bear, squatting and pointing at the ground. "Looks like an armored division has been here. Tank tracks everywhere."

"Theirs or ours?" asked Cardini.

"Must have been ours," answered Duthie. "There's a spent pack of Raleighs."

"Who the hell you think came up with this little gizmo?" Chicago asked, staring at the P38.

"Who the hell cares?" responded Averitt.

Cardini shook his head sarcastically. "I'm sure it was some pinhead college boy who drew up some plans one day, sold fifty million to the Army, and right now is sitting in the lap of luxury back in the States with his little exemption."

"And is surrounded by women," added Chicago.

"And lots of good food," added Spotted Bear.

"And lots of women," added Chandler again, who had lost interest in his meal and was now in the process of brushing his teeth.

I fastened a clip onto my rifle and aimed at a pinecone high in a tree in front of me. I squeezed the trigger, and the pinecone exploded with a crackle. A puff of smoke went up from the spot as tiny remnants floated down to the ground.

"Lookee here," Spotted Bear said, aiming. "The little one just to the right of that." He shot. The smaller one broke into a thousand pieces as the branches of the tree shook.

"My turn," challenged Duthie before he pulverized a big one high atop the next tree over.

The shooting gallery went down the line a couple of times. The tops of the trees across the way were reverberating from the volley of gunshots when Sergeant Thomas came running in our direction. "What the hell are you boys doing?"

"Just shooting at those pinecones," I answered, taking aim at my next victim.

He put his hand on the top of my carbine and pushed it to the ground. "What the hell's wrong with you, Sacco? What's the matter with all you guys? Don't you know you could be giving away our position to the enemy?

"This ain't the county fair!" he continued. "You ain't gonna get a stuffed animal for hitting a damn pinecone in a tree! Okay, playtime's over. Put up your toys, and let's get to work. Come on! We got a job to do."

Maybe it was because it was the first semiquiet day we'd experienced in a long time and we needed to blow off some steam, or maybe it was because, underneath it all, we were still a bunch of teenagers with more enthusiasm than brains, but, for whatever reason, it didn't occur to any of us that a little target practice might conjure up any adverse affects. Sarge was shaking his head, looking around in disbelief. "Damn! You would have thought that after all the training I put you through *to save your own damn lives* that it would have taught you something . . . *something . . . anything!*"

We put away our weapons, picked up our tools, and went to work. A few of the guys shimmied up trees and began tying off lines from the large spools of cable on the back of each truck. Duthie and I climbed atop our truck onto a small platform, which Tex then jacked up an additional five feet to the lower branches of a large tree.

We were minding our own business, happily working away, when we heard the whistle of a German bomb coming at us. The explo-

Silverman (left) and I string cable in France, 1944.

sion sent shock waves through the trees. Another missile followed in hot pursuit behind the first. This one was closer. Then a third. And a fourth.

The truck was rocking below me. Duthie was lying down on the platform, trying to crawl slowly backward onto the roof of the bouncing truck. More bombs were coming in. I turned to reach for the line I'd been tying to the tree, hoping it would give me some stability in case the truck flipped out from under me.

As I reached out, I caught sight of a man flying toward me through the air. I looked in his direction just in time to realize it was Sam Martin, swinging on a wire he had fastened high up in a tree, picking up speed as he approached, screaming at the top of his lungs. Behind him I could see trees splitting into flames as incoming bombs exploded ever closer. I did my best to dodge him, but he clipped my left leg as he passed and suddenly I was airborne.

I landed on the ground, rolled a few times, and then tried to get up and run. The ground was shaking like jelly, and the truck was dancing. Guys were jumping out of trees and racing for cover. Duthie was still half on, half off the platform, embracing it as it bounded up and down, riding it like a bull at a rodeo. Averitt, Cardini, and Spotted Bear were under a jeep with their heads covered.

German fighter planes zoomed just above the treetops, spraying a hail of bullets downward through the leaves. The rounds kicked up dirt, plugged trees, and ricocheted off the trucks in every direction. I ran a few steps but fell as the concussion from a blast jolted the ground. Chandler ran past me, then turned, came back, and, in one sweeping motion of his arm, lifted me to my feet. I followed him toward a five-foot-high stone wall near the edge of the trees as a trail of bullets headed in our direction. Chandler launched himself at a small hole near the base of the wall. I jumped up and flipped over the top. When I landed on the other side, Chandler was there, crouched against the stones. I could hear the bullets hitting the wall and then the field beyond us as the planes flew over.

The raid lasted several frightening minutes. I stayed in that position until all was quiet. Finally I stood and looked over the wall. Guys were stepping out of their hiding places. Two trucks had been hit, a jeep overturned, and a few trees broken in half and smoking. The place was shot up, but nobody seemed to be hurt.

"Give me a hand up over this damn thing," Chandler said, reaching for the top of the wall, hoisting one leg up in the air toward it.

"You can make it," I said, lifting myself up and flipping to the other side.

"Don't you leave me over here, Sacco," he begged.

I could see Chandler's hands and the top of his bald head as he repeatedly attempted to pull himself up, only to fall back down. "Hey, Duthie!" I yelled. "Come help Chandler. He can't get his fat ass up over this wall."

"How the hell did he get over there?" Duthie asked, still dusting himself off.

"I came through the hole," Chandler responded.

Averitt, who had wandered up, looked perplexed. "Well, then," he said. "Come back this way through the hole, dumb-ass!"

"I can't. I can't fit."

We bent down to see the opening. It was about a foot and a half at its widest point. "Hey, Chandler," Averitt called out, "how the hell did you get through this thing?"

"What do you mean, you can't fit?" I asked. "You went through it that way."

"What's the problem here?" Sarge asked as he approached. "Everybody okay?"

"Yeah. Chandler's on the other side of the wall," Duthie answered.

"Okay," said Sarge. "Get back over here, Chandler. Let's get finished up here and move out before they come back."

"Hey, Sarge, let me ask you something," Averitt said. "How the hell did Chandler fit his fat ass through this hole?"

Sarge looked down at the aperture. "Chandler, I thought I told you to get over here."

"I'm trying, Sarge."

"Come through the hole."

"I can't, Sarge. I tried."

Sarge looked at Averitt and me. "Help him."

We climbed up on the wall and each gave Chandler a hand. He pulled himself up, sat on top, swung around, then hopped down to the ground.

"Now, tell me something," Sarge said. "You went through this hole, but you couldn't get back through it."

Chandler studied the hole as if he were seeing if for the first time. "Yeah, I guess not," he said.

Bird Turd, who was about half the size of Chandler, was walking near us. "Wilkins, get over here," Sarge ordered.

"What?" he asked, coming closer.

"We got a special mission for you, Bird Turd," Averitt said.

"Wilkins"—Sarge was pointing at the wall—"get down here and see if you can squeeze through that hole."

"He might get stuck," warned Duthie.

"Good," replied Averitt.

Bird Turd looked around at us, got down on all fours, and poked his head through the opening. He pulled it back out, took off his helmet, and then stuck his head back through. He wiggled and wrestled and contorted until he popped out the other side of the wall.

"Would've been easier just to jump over the damn thing, Sarge," he said. "What do you need me to do over here?"

"Let's leave him," Averitt suggested.

"Come back over," Sarge called out. "Chandler, I'd like to have seen you get through that damn thing."

Chandler laughed. "Well, it was either get through that hole or get a set of bullets up my ass." He turned and started walking back toward the truck. "Come on, Sacco," he said. "We got work to do."

We would occasionally stop to answer nature's call as we pushed along. Captain would halt the trucks for a couple of minutes, we would jump off the back, walk to the path's edge, line up side by side, and pee. Looking down the line, you could see streams of water shooting off the side of the road like a giant fountain, with each man trying to outdistance the other. As a particular stream died down to a drip, that man would repack his hydrant, jump back on his truck, and be ready to go. The whole process didn't take longer than a couple of minutes.

We happened upon a series of stopped railroad cars late one afternoon, their noisy engine adding water from a wooden tank, their passengers apparently taking a piss call. Fifty or so soldiers were in the tall grass beside the tracks as we approached. Captain waved for our trucks to stop.

"These are Russians," said one of our defectors who recognized their uniforms across the distance.

Averitt pulled out some binoculars and took a closer look. "Oh, my God," he said. "They're women!"

We all scrambled for the binoculars as one of our Russians yelled out something to the women. They looked in our direction, screamed, and ran toward their train. When they'd boarded, most of them looked back at us, pointed, and laughed.

"Come on, boys," Sarge said. "Let's keep moving."

Averitt was still looking through his spyglasses. "That was the damnedest thing I ever seen."

"What's the problem, Averitt?" Cardini asked. "You never seen a woman pee before?"

"Hell no," Averitt replied. "Not standing up!"

I was getting good at poker, at least good enough to win from these guys. It's not as if I could go to the store and spend any money I won, but it was better that it be in my pocket than in theirs. Besides, winning a card game or two helped me feel like I was in control of some small part of my universe as the chaos roared all around.

In the late afternoon of August 16, after thirty-six hours on the move, we stopped in the forest outside a town called Longny. One of the food trucks had met us there, so we set down our packs, grabbed our mess kits, and lined up for hot chow. After dinner I was so tired that I felt as if I were about to implode. But I had just enough energy for one or two quick hands of poker.

We spread out a blanket and got busy. The other boys must have been more tired than me, because I won a few hundred dollars. As I pulled my winnings toward me, I heard a voice say, "I'll take ten percent of that."

"Like hell you will," I responded, a cigarette dangling from my mouth. I turned to find a captain—who also happened to be a Catholic priest—standing behind me. I took the cigarette from my mouth and coughed. "Uh, sorry, sir . . . uh, Father. I didn't know you were . . ."

"Nah," the priest laughed. "I was just kidding."

"I'm out," I said as I tossed the cigarette, gathered my money, and got up.

"Hold on, Sacco," Hutter said. "You can't leave until we win our money back."

"I think I'll quit while I'm ahead."

"Have a seat, Padre," invited Hutter.

"Yeah, pull yourself up a piece of blanket and get in here," added Chicago.

"Well, maybe just one hand," replied the priest.

"Yeah, and I hope you're not as lucky as Sacco there," said Cardini.

I almost said, "Ah, go to hell," but thought better of it. I walked over and got a cup of coffee, then came back and sat against a tree. I lifted the drink to my face, smelling its aroma, feeling its warmth on my fingers as I held the cup with both hands. I took a long swallow. It felt better than it tasted, but it was still better than nothing. I set the coffee down and rummaged in vain through my backpack for a piece of paper and a pencil.

The other guys were still playing poker on the blanket. It looked to me like the priest was winning. Most of the trucks were parked in an area to my right, where Tex and a few of the other drivers were, as usual, tinkering with them. A small group of guys was having an animated conversation off to the left, but I couldn't make out what they were saying. Averitt and Duthie were walking, smoking, and talking about something. Many guys, including Silverman, were sitting around quietly writing letters. Some were cleaning their weapons. Others were asleep on the ground. In the distance all along the horizon, I could see flashes of light as the battle continued. I stuck my hand down into my backpack again, shut my eyes, and felt around for that pencil.

Softly, almost imperceptibly at first, I heard birds singing. The sounds were sweet, innocent, peaceful, and unexpected. I caught the slightest whiff of herbs carried on fresh air and bread baking nearby. I slowly opened my eyes to see the sun filtering through the leaves of the trees and dancing on the ground around me. The guys and the trucks were nowhere to be seen. I was a bit confused, so I stood and picked up my rifle and backpack.

In the distance I could faintly make out the strains of Italian songs being played on mandolins, being sung by familiar voices. As I walked around the giant oak tree, I was amazed to see Grandpa Sacco walking toward me. Farther off I could see Papa, Uncle Vincent, Uncle Joe, and the others coming in from the fields. To the right, near the place where we'd been playing poker, my grandmothers and aunts were setting a large table and filling it with food. My little cousins were running around in all directions, circling the area, and eventually homing in on the table and the feast.

Grandpa Sacco had reached me. Now he held out a small package, wrapped in a white cloth and glowing from within. I looked carefully at the gift, trying to discern what it could be. He held it toward me and smiled. And I accepted it.

He then turned and motioned at the green fields with his hand. The sun was bright and warm over the landscape. A gentle breeze caressed the rows of corn and tomatoes. They swayed in surrendered unison as the calming waves of summer air slowly moved among them. A set of chimes, hanging on the porch and rustled into action by the wind, dutifully played its quiet notes.

The screen door to the kitchen slammed shut as Grandma Amari came out carrying a beautiful roast. Suddenly she dropped the pot, which bounced on the ground with several loud thuds. There was an inexplicable noise coming from the field to the left, almost as if a tractor had fallen in a ditch, creaking and moaning as it plummeted down. Papa and Uncle Vincent jumped to their feet to see what was going on. Both knocked over their chairs in the process. A couple of the uncles began yelling, and several of the younger children, frightened by the sudden commotion, started crying.

The screen door slammed again. I looked around in time to see Mama running toward me from the direction of the house, a frantic look on her face. "Joe! Watch out!" she yelled. "Get up! Run, Joe! Run! JOE!"

As she closed in on me, I heard a massive explosion behind her. I was startled into the present.

"JOE!" I heard Duthie yelling as he ran past me. Loud explosions were lighting up the campsite. "Move it, Sacco!" Averitt added as he blurred through the scene. "Run!"

I grabbed my rifle and my pack and started running after Duthie and Averitt. They jumped into a small ditch at the edge of the camp. I jumped directly on top of them as a mortar hit the tree against which I'd been sleeping, cracking it in two, blasting jagged, burning shards of lumber out in all directions

We kept our heads buried for the duration of the attack. Once the bombing subsided, we peeked out at the condition of the camp. Guys were slowly rising from trenches and foxholes, gingerly walking about, checking the extent of the damage.

"It just never fucking stops, does it?" Averitt asked rhetorically as we climbed out of the ditch.

"It stops when one of 'em hits you," I answered.

There was a huge, burned hole in the tree just above where my head had been. "Uh, thanks for waking me up, Duthie," I said. "I owe you one, buddy."

"I'm sure you'll have plenty chances to return the favor," he said as we continued walking.

"Awe, sheeeeiit, Sarge!" we heard Tex almost crying up ahead. "Look at my goddamn truck! Son of a bitch!"

Sarge and Captain English were approaching the scene. The truck had taken a series of direct hits and was contorted, bent, broken, busted, mangled, and mutilated. And it had at least six flat tires.

Sarge studied the vehicle, then looked back at Tex. "Can you fix it?"

"Awe, sheeeeiit, Sarge! Sheeeeit!" Tex was staring at the truck as if it had been his best friend.

Sarge raised his shoulders and looked around at the captain. "Just get him another damn truck," English said.

The bombardment resumed and continued through the night as the battle escalated around us. The American artillery launched a tremendous barrage of heavy shelling against the enemy, lighting up the night sky with loud, spectacular, and deadly fireworks. Each side was bombing the other with such ferocity that the trees shook and swayed above while the earth trembled beneath.

We hunkered down in our refortified camp, with trenches deep, armament up, and weapons cocked, as the two magnificent heavyweights slugged it out with increasing power and determination. Silverman explained that the Germans didn't know how to aim their big guns, that if they were trying to hit us, they would end up hitting something a mile or two away.

"But . . . what if they're aiming at some artillery unit a mile in that direction?" Cardini asked.

"Then run like hell," Silverman replied.

With the passing of each hour, the explosions seemed to be moving slightly farther away. We were giving the enemy some hell, and they seemed to be retreating, if ever so slowly.

Amid the terror of combat, I found myself thinking more and more about home, of what it used to be like and what must be happening there now. Sometimes I'd look at my watch and try to figure what time it would be back in Birmingham, so I could gauge if they would be eating or sleeping or going to the market or maybe just sitting beneath that oak tree, drinking wine and talking. I wondered if they were wondering about me. I wrote lots of letters but didn't know how long it would take them to get there. That, I assumed, depended on what was going on around here.

And there was plenty going on around here. Not only were our telephone lines continually being destroyed, but the infantry was moving so fast that we could barely string new wire fast enough to keep pace with them. We were working desperately to keep the communications up and running as we moved steadily in the direction of Paris.

We set up our headquarters near the town of Dreux, alternating to the front with Company B, working ten or fifteen hours, getting shot at, then pulling back into friendlier territory for some rest. The locals in this area had offered us use of their houses and barns, and we had enthusiastically accepted. They liked the idea of having American soldiers on their property. We liked the idea of having better shelter than a pup tent and better food than a K ration.

Duthie, Chicago, Chandler, and I were walking down a dirt road late one afternoon, returning from a work detail near the front lines, when we saw Silverman running toward us.

"Sacco, come quick!" he was yelling. "Guys, hurry!"

"What the hell is this?" Chicago asked. "Whatchu need, little Jewish man?"

Silverman was out of breath by the time he reached us. "Copeland tried to rape a French girl, and Spotted Bear and Sam Martin are gonna kill him!"

"Oh, shit!" Duthie said as Silverman changed his course, and we took off running after him.

Copeland was a black soldier who had been transferred into the 92nd while we were in England. He had met with the normal resistance afforded any replacement, but because of his skin color and the general segregation of the military at the time, his acceptance was somewhat less than enthusiastic. To make matters worse, he had been assigned to share a tent with Tex, who wasn't exactly known for his tolerance or pleasant demeanor. Needless to say, the living arrangement didn't sit well with either of them.

After a great deal of consternation, mostly instigated by Tex, Sarge had gathered everyone together and, holding up an official-looking piece of paper, announced that there needed to be no further concern about the issue because, as he put it, "Copeland's medical records came in today, and they say he's tested mostly white."

His medically tested whiteness—whatever that meant—was reason enough for most of the guys to ignore the novelty of his race and treat him with respect. In fact, once we got to know him, Copeland turned out to be a nice enough guy. Still, because of the inevitable clash of cultures inherent with any invading army—and the availability of alcohol—the situation remained somewhat delicate.

We found them beside a barn, Copeland on the ground, Sam Martin twisting his arm behind his back to keep him immobile, Spotted Bear sticking the business end of a carbine in his mouth. All three were dirty, bloody, and beaten. And drunk.

Cardini and Hutter were watching helplessly. Sarge was sprinting up the small hill from the road. "Spotted Bear! Let go of that soldier!"

The lady of the house was crying and shouting something in French at the men as she embraced her teenage daughter, who was staring doe-eyed at the spectacle.

"I said to let go of that soldier, goddamn it!" Sarge was hollering in Spotted Bear's face.

The French lady began yelping at Sarge, waving her hands around, pointing at Copeland.

"Let him go!" Sarge demanded.

"Spotted Bear!" I yelled.

"None of your business, Sacco," said Sam Martin. "Keep out of it."

"Spotted Bear," I continued, "they'll lock your ass up and send you to the electric chair!"

Spotted Bear gritted his teeth, set his jaw, and readied himself to fire the weapon.

"What the hell's wrong with you?" Chandler asked.

"Don't do it!" I screamed.

"Let him go so I can beat the hell out of him," Chicago added.

"Let him go," I repeated.

"Take that weapon out of his mouth," Sarge commanded. "Or I can guarantee you that I will have you court-martialed and sent to prison for the rest of your life for killing an American soldier!"

Spotted Bear became completely still and was staring down at his captive.

"Go ahead and kill him," Cardini said. Copeland's eyes widened. "Shit, what do I care?" he went on. "What do any of us care? Kill the son of a bitch. Then they'll send you to prison and kill your ass. What the hell do we care? After all we've been through, you just want to throw it all away by killing that asshole, then go ahead. Let him shoot him, Sarge."

Sarge turned to Cardini. "Look, I don't really think we need any of that reverse-psychology shit right now, do we?"

Chandler looked at Cardini, shook his head and rolled his eyes. "College boy."

"Just take the gun out of his mouth," Sarge instructed again.

"Spotted Bear," I said calmly. "Spotted Bear." He took his eyes from Copeland and looked up at me. "Let him go. Sarge will take care of him. The Army will lock his ass up, don't worry."

He grimaced a bit, then yanked the rifle from Copeland's mouth, falling backward. Chicago and Chandler ran to confiscate his carbine and help him up. Sam Martin released Copeland's arm with a push, smashing his face in the dirt one more time. Sarge grabbed Copeland and pulled him roughly to his feet. The French lady was closer now, flailing her arms about, jabbering away at Sarge.

"Now, somebody tell me what the hell's going on here," Sarge demanded. He looked around at the lady. "Somebody other than her."

"This nigger here," Spotted Bear began, pointing at Copeland, "was telling that woman that he's an Indian."

"What?" Sarge asked.

"He was raping the girl, Sarge," said Sam Martin. The lady, emboldened by our restraining presence, wagged her finger in Copeland's face.

"I didn't rape nobody," said Copeland.

"But you *tried* to rape her, right?" Sarge clarified.

"No, First Sergeant Thomas," he answered. "That ain't right. I . . . I . . . I tried to *kiss* her."

"You're a damn liar," Spotted Bear slurred, struggling to get away from his captors.

"I was just gonna kiss her, First Sergeant. That's all," Copeland said as he wiped the dirt and blood from his lip.

Sarge turned to the lady and the girl. "What happened?" he asked slowly.

The lady began talking fast, pointing at Copeland, pointing at her daughter, then at Copeland, then at Spotted Bear, then at the girl again.

"Hold up, hold up," Sarge said. "Anybody understand what she's saying?"

Silverman jumped in. "I think she's saying something . . . about Copeland . . . and he . . . did something . . . and . . . this girl . . . and something about Spotted Bear."

"That's just brilliant," Chicago said. "Thanks for your translation services, you idiot." He turned to the lady and yelled, "HE . . . RAPED . . . YOUR . . . DAUGHTER," very slowly and very loudly in English.

The lady looked confused. I decided it couldn't hurt to ask her what had happened in Italian. We certainly weren't getting anywhere in English or French. We went back and forth a few times, but I was finally able to make out that Copeland had wandered up and asked the girl and a cousin of hers to go on a walk. When they refused, he grabbed this one and forced himself on her. The girls screamed, and this one's mother ran out of the house, yelling and hitting Copeland.

"What's she saying, Sacco?" Sarge asked.

"I think he grabbed that girl, and she started screaming," I answered.

Sarge turned to Copeland. "Is that what happened?"

"Well, yeah, but—"

"And we were walking up that road, and we heard this lady hollering some god-awful shit in French and hitting this nigger, and those girls were crying," Sam Martin piped up.

"Okay, that's enough," Sarge said. "Say 'colored' or 'Negro' from now on. Understood? Now, there's more than one girl?"

"And when we got up here, he was telling that French lady there that he's not a nigger, that he's an Indian!" Spotted Bear was as incensed as he was inebriated. He turned and said loudly to the lady, "He ain't no Indian! He's just a nigger!"

"I told you that's enough with that word, damn it," Sarge snapped. "Where's the other girl?"

"Dov'è l'altra ragazza?" I asked. The lady pointed back to the house.
Sarge turned again to Copeland. "You trying to rape these girls?"

"No, sir, Sarge. I just tried kissin' one of 'em."

"Don't call me 'sir.' All right . . ." Sarge looked around at Cardini
and Hutter. "You two, help me take this one back to see Captain
English." He pointed at Spotted Bear and Sam Martin. "You're
coming, too."

"And then I came walking up the road there, and I heard all this
screaming and—" Silverman began.

"He was telling her that he's an Indian, Sarge," Spotted Bear interrupted. "He ain't no damn Indian. You ain't no Indian, you big, ugly,
black-assed nigger!"

"Shut the hell up!" Sarge ordered.

As they led the three away, Silverman tried to finish his version of
the story to anybody willing to listen. "I was coming up the damn
road, and I heard all this racket."

In the distance I could hear Spotted Bear repeating, "Damn it if he
ain't a son of a bitch, and he *ain't* no Indian, Sarge. *He ain't.*"

Captain English detained Copeland and brought a translator to
talk to the woman and the two girls the next morning. I guess what
Copeland had said was right—he was only trying to kiss the one girl
when her mother ran out of the house in a frenzy.

Captain said he didn't need the aggravation anymore, so he reassigned Copeland to what was known as the Red Ball Express, a division of black soldiers who drove a nonstop convoy of supply trucks
from the ports on the coast of France to the front lines and back
again. He was on a jeep heading west by noon.

Nothing much really happened to Spotted Bear and Sam Martin.

Later that same afternoon, our new truck arrived from Cherbourg.
Tex said it turned out to be the happiest day of his life.

13

Paris and Beyond

August 19, 1944

Of special importance in the war being waged by the boys of the 92nd Signal Battalion was getting to the next town before the infantry so that we could be the first to find any grateful-to-be-liberated girls. Sometimes we would let the infantrymen fight up to the edge of a town, and then we'd go in ahead to start setting up phone lines. It wasn't the safest way to operate, but the prospect of discovering pretty girls made us more than willing to take the risks.

Each town was fantasized to be the dream town . . . the town in which lived nothing but beautiful young Frenchwomen, all anxiously waiting for the day they would meet a handsome young American soldier and be swept off their feet. Most towns did have their share of beauties, but they were always mixed in with ugly people, old people, fat people, and kids.

The next town, though, held great promise, for the next town was Paris.

We came upon the Seine River on August 19, 1944, near the town of Mantes-la-Jolie, a few miles west of Paris. A group of us emerged from a heavily forested area, where we had set up a communications board for Company C, out onto a rocky bluff overlooking the river valley. A thick stand of trees created a hidden, protected balcony from which we could witness the action below.

The stately river transported itself slowly through the valley, turning slightly this way and that to accommodate the landscape, stretching forward until it disappeared around a bend too far away to see. The nearby hills had a few outcroppings of granite rocks near the tops but were completely covered with the green of trees and shrubs below as they bowed to the wishes of the river and became the valley.

A fierce battle was being fought for a bridge below us. By late afternoon the Americans had emerged victorious, destroying German positions and vehicles near the bridgehead and clearing out booby traps underneath. The infantry was sending the first convoy of trucks across the Seine when two enemy fighter planes appeared from over the hill to our left. They swooped down into the valley and began strafing. The trucks stopped, and the soldiers ran for cover.

In an instant a barrage of Allied antiaircraft fire went up from all over the valley. To my surprise one of the big guns was only about twenty yards from us, concealed in the thickets, now blasting away at the planes. The skyward storm continued until the Nazis raced away from the valley and into less hostile environs.

Once the Germans had departed, the Americans climbed back out onto the bridge and continued the crossing. An increasing amount of matériel was being transported over the structure when we saw five American P47 aircraft flying up the valley in our direction.

"Here come the good guys," Chicago said.

"Should have been here a little while ago when the krauts were buzzing the place," Chandler added.

They were coming in low through the hills, and the guys in the area of the bridge were watching them, waving, giving them the thumbs-up.

"Something's wrong," Silverman said, squinting through his thick glasses at the planes.

"Yeah, what's that?" I asked.

"The Army Air Force flies in pairs or fours," he said, still staring into the sky. "There's five of those suckers!"

As the planes neared the bridge, the third and forth ones swooped down and unleashed a torrent of bullets on the startled soldiers. The fifth came even lower and leveled two bombs at the bridge itself. Both overshot it and exploded in the river a few hundred yards downstream, sending huge plumes of water into the air.

Caught unaware and now frantically swinging into action, the antiaircraft guns concentrated all their firepower on the fifth attacker. As it cleared the bridge, the plane caught a host of Allied gunfire and burst into flames. It careened through the air, wings dipping back and forth, smoke billowing from the tail section and one engine, until it crashed into the riverbank. It exploded into a red ball of fire. A cloud of black smoke rose above the burning trees.

The American planes had apparently been captured by the Germans, who were using them—and the element of surprise—against us.

Sergeant Thomas called for us to get across the bridge before bad things started happening again. There were a few more attempts by the Germans to strafe the area, but the antiaircraft gunfire kept them at bay until the real American flyboys finally arrived and cleared the skies. A few random shots were fired at the bridge from hidden spots throughout the valley, but nothing too serious. With the help of the Army Corps of Engineers, we managed to pull cables across the Seine and establish the communications that would take our army to Paris and beyond.

Though we were anxious to enter the City of Lights, orders came down for us to swing south around the city. The American government wanted the French army to have a major role in the liberation of Paris itself, so only specific U.S. personnel were being allowed in town. Many of the guys were upset. We felt as if the French army, through its own ineffectiveness, had played a major role in allowing the Nazis to take over France in the first place. Furthermore, as we were fighting through *their* country to liberate *their* people, *their* military was either getting in our way or nowhere to be seen. But mostly we were upset because we wanted to get into Paris and find some women.

The outlying towns were just as happy to see us. People were in the streets, shouting, singing, and thanking us for what we had done so far. But all was not festive. A group of men rounded up several Frenchwomen who had collaborated with the Nazis. They stripped the women, shaved their heads, and painted swastikas on their breasts. They loaded the women onto the backs of trucks and paraded them through town as the locals jeered and berated them, hurling insults, rocks, and bottles at them as they passed. We witnessed this same ritual in several of the small towns around Paris.

We asked one Frenchman what would happen to the women, and he replied, "They will get what they deserve." We watched the townspeople escort the women out of sight, but we never learned of their fate.

Fontainebleau, France
August 26, 1944

The morning was bright and beautiful, and the streets of Fontainebleau were lined with people welcoming us with cheers, hugs, and flags a-waving. A military band was playing some French music, but it still sounded good. Most of the soldiers were riding on the tops and the hoods of vehicles as Tex and the other drivers blasted their horns and did their best to see around us. The tanks, trucks, and jeeps were being covered with flowers tossed by grateful citizens. Averitt, Chandler, Spotted Bear, Silverman, and I hopped from the truck and started walking beside it after we noticed girls in the crowd running into the street to hug and kiss the soldiers. In addition to the quick smooches, they handed us more flowers and bottles of wine.

Some of the infantrymen up ahead were tossing inflated balloons into the crowd. The people were bouncing them back and forth to the soldiers as the merriment continued.

"Look at that," I said to Cardini, pointing up ahead. "Somebody brought some balloons."

Silverman was straining to see. "I haven't seen balloons since I was a kid."

"You think those are balloons?" Cardini asked. "You guys are amazing! Hey, Averitt! These two choirboys think those are balloons bouncing around up there."

About that time one of the "balloons" bounced right into Silverman's face. "Oh, shit, would you look at that, " he said, grasping the toy. "It's a rubber!" He hit it toward Cardini.

"Yeah," Cardini replied, catching it. "And it's Army-issue, so if it doesn't work, you can just blame Uncle Sam."

Silverman ran up to take another look at it. "It ain't Army-issue. Is it?"

"Yeah, Silverman," Averitt answered. "Didn't you get your stock from the supply sergeant back in England?"

Silverman grabbed my arm. "You guys are shitting me, aren't you?"

As we walked along, I spied an attractive young lady standing on a balcony with what seemed to be her parents, grandparents, and younger brother. "Look up there," I said to the guys. The girl smiled and threw a flower as her family clapped in time with the band on the street and waved French flags. I made a mental note of the address.

After we established our campsite on the outskirts of town, I convinced a few of the guys to go back into town with me to find the girl. It didn't take a lot of convincing.

We found the address without a problem, but when an elderly man answered the door, he claimed to know nothing of a young woman who had been there. It was the same old man who had been on the balcony with the girl that morning.

"I saw her right out there," I told him. "She waved to me and threw me a flower."

The old man shrugged as an old woman entered the room.

"I just wanted to thank her, that's all," I said.

He said something quietly to the lady. "We don't have any young girl here," she said to me.

"No," I insisted. "This is where she was, because I saw you up here with her. I remember you."

"No, you make the mistake," she repeated. "It is easy with all of this . . ." She made a motion to indicate the hoopla that had taken place earlier on the street below.

The man was tougher to understand, but I believe he said, "We offer you and your friends a drink, and then you go."

I was about to give up when the old man threw his hands in the air and the woman said, "No, please no!"

A man I assumed to be the father of the girl stepped out from the kitchen, looking quite alarmed. He held his hands up in a calming gesture. "Please," he said, his eyes wide and darting around the room. "We . . . we will let you talk to our daughter, but please do not hurt her or us."

I looked around behind me to find a drunken Cardini in the very back wielding a pistol over his head. "What the hell? Put that damn gun down, you stupid son of a bitch!" I yelled. "What the hell's wrong with you?" I turned back to the frightened family. "I'm sorry about this," I said, backing out of the doorway. "Chandler, get that damn gun away from him."

Just before the old French grandfather shut the door, I caught a glimpse of the girl peeking around the corner from the kitchen, smiling. She was a beauty.

Other, more serious matters emerged as we walked back to camp. Many locals came up to offer us drinks and thank us. Most wanted to tell us what had happened in their town before we had arrived. They told stories of German atrocities that were quite disturbing—local men castrated in public for providing information to the French underground resistance, a woman beheaded for not cooperating with the Gestapo, a Nazi walking through town at sunset with a baby impaled on a bayonet, the Germans dragging the bodies of three American POWs through the streets behind a jeep. The stories went on and on. They would have been hard to believe had it not been for the passion in the townspeople's voices and the tears in their eyes.

They embraced us and kissed our hands. They were free again.

The Germans were retreating quickly now, and we were on their tails. We moved north to Rozay-en-Brie, then south to Sens, then east to Troyes and Neufchâteau, crossing the Seine several times as we placed and repaired lines for the advancing infantry. Though we had the enemy on the run, the sounds of battle still surrounded us day and night. On rare occasions everything would get momentarily quiet. Guys would look around, wondering what was wrong. Before too long the bombing would start up again, its concussive noises now strangely reassuring as they rumbled like waves through the ground.

We had the feeling that we had whipped the enemy and that the war would soon be over. But even as we moved deeper into France, I found that I was still haunted by images of what I'd seen in Normandy. I kept thinking of all the boys who had died there, of that one young fallen soldier I'd stopped to see, and of all the others I'd seen dead since then. I would often say a prayer for them and their families, and I would pray that God would keep me and all of my buddies safe.

Mail call was our daily miracle. Even in the midst of action, writing and receiving letters was a top priority, right up there with finding girls. I, like most of the guys, would read mine a thousand times, memorizing them and then putting them away, only to pull them out again whenever I missed Mama or Papa or one of the cousins. It was comforting to see the handwriting, to know that someone back home was thinking about me.

The letters also brought us into each other's lives in a more personal way. The birth of a new niece or nephew for one of the guys was greeted with congratulations and reminiscences of our own families.

Me reading a letter from home in France, 1944.

Word from a wife or girlfriend was envied. A joke or funny story was read aloud, shared with the group so that all could, for that one moment, feel connected to something other than the misery surrounding us. In a very real way, good news for one was good news for all. Still, not everything was for public consumption, and it was not uncommon to see a guy wiping a hidden tear as he read or reread a cherished letter from home.

In addition to my own letters, I was still reading and writing all of Chandler's. After a while I probably knew more about his life than he did. One thing he didn't have trouble understanding was a picture of Betty Grable sent to him by his uncle. He kept that folded up in his helmet and would occasionally take it out, stare at it for a few minutes, then put it neatly back in its place. And every now and again, Silverman would ask me to write a missive from him to Rosanna. He, unlike Chandler, could read and write for himself, but he said that he just liked the way I phrased stuff. I didn't mind.

As it had done since our days in basic training, the Army was censoring all mail coming and going so that sensitive information would not fall into the wrong hands. If a letter contained too many details concerning a plan or operation, that part of the letter was cut away and the remainder sent. Opinions were allowed, but facts and figures were discouraged.

Despite a prohibition on the disclosure of our whereabouts, I decided to encode a message to Uncle Paul back home. At the end of a rather mundane letter, I said, "Tell all the cousins I said hello, especially Cousin *Frances*." After a few weeks, I received an answer from him in which he wrote, "I don't know who you're talking about because we don't have a cousin named Frances."

Near Châtenois, France
September 5, 1944

We were walking across a field, heading for the trucks at the far end, when I heard a crackling sound beneath my foot. The ground gave way, and in the next instant I fell into the earth. I only had time to make a short yelping sound as I went down. A split second later I heard another, louder yell as Chandler crashed down beside me.

"Sacco!" Averitt called out.

"Chandler, you okay?" I asked.

"Hey!" Silverman was poking his head through the hole. "Hey, you okay down there?"

"Just quit playing around, and help me outta here," I said, reaching up.

"It's a trench," said Chandler.

"He needs a wrench," Silverman called back to Duthie.

"Get . . . just . . . look, just move out the way," I heard Chicago saying as he pushed Silverman back.

"Whatchu doing down there?" Averitt asked.

"I fell in this damn hole," I answered. "Chandler's—"

"Yeah, I know," Chicago said. "Chandler fell in, too. I heard."

The guys up top moved away the twigs and branches covering the hole to reveal a trench probably left over from World War I. It was about five or six feet deep, about twenty feet long, and mostly overgrown with grass and mold. It appeared to have been recently occupied and camouflaged by someone who'd made a hasty retreat from the area. Left behind were a pocketknife with a German inscription on the blade, what looked like a military belt buckle, a few spent rounds from a rifle, a smattering of cigarette butts, and, most intriguing of all, a case filled with phosphorus grenades.

We climbed out of the hole, loaded the grenades into the truck, and got on about our business. The infantry was preparing for a major push into the French region of Alsace-Lorraine, the final stop before our invasion of Germany. And wherever the infantry went, so went the signal battalion.

We were sitting on the back of the truck watching the landscape move away as we rolled eastward through the hills toward a little town named Mirecourt. In the western sky, we could see the late-afternoon sun reflecting off the wispy clouds. "Look at that," Averitt said. "Looks kinda peaceful, don't it?"

I pondered it for a second. "Looks like a painting."

"There's this guy back home, he's from Elgin . . . well, near Elgin . . . well, I don't know where the hell he's from, but he's a great artist," Chicago was expounding. "Son of a bitch can draw anything."

"Yeah," Cardini interjected. "I got an uncle like that who can draw anything. And can play the piano, too."

I had opened the case of German grenades and was rummaging through them.

"No, I mean this son of a bitch can draw *anything*—like a real old-timey artist," Chicago continued. "Animals, people, clouds like that, anything. Fruit . . . bowl of fruit . . . make you want to reach in there, grab a piece, and eat it."

"Sounds like he might be one of the great artists of the world," Averitt said sarcastically.

I picked up one of the grenades and studied it carefully.

"You guys think I'm shitting you, don't you?" Chicago said. "But I'm telling you, this guy's . . . well, he's as good as, if not *better* than, Michelangelo and Leonardo da Vinci and all those guys."

"He's not better than Michelangelo," I clarified. "Nobody is." With that I tossed the grenade off the back of the truck. It hit some rocks in a small ravine and exploded, sending up a white, glittering plume of fire and smoke.

"You see that damn phosphorus?" I asked excitedly.

"Holy shit!" Cardini said. "Give me one of those." I handed him one.

"What the hell are you talking about, Sacco?" Silverman asked. "Michelangelo was one of the great ones, but he's not the greatest of all time."

"Hard to beat Michelangelo," said Averitt.

Cardini threw his grenade. It exploded with a bright, round, sparkling light.

"Looks like fireworks," Duthie said. "Here. Give me a couple."

"Well," continued Silverman, "what about Rembrandt?"

"What about him?" I asked.

"You saying Rembrandt was a better artist than Michelangelo?" Cardini asked before pitching his next one.

"I don't know," Silverman began. "I'm just saying there are some art critics—"

"Hey, let me tell you something," I cut him off. "Rembrandt don't amount to a pimple on Michelangelo's ass."

Duthie hurled two into the trees. They exploded one after the other. Not to be outdone, I quickly launched three high into the air. In fact, I launched them too high. They didn't make it to the woods or the ravine but exploded in the road, sending up three spectacular orbs of shimmering light, smoke, and dirt.

Before the haze could clear, a jeep sped around the hill in our direction. The jeep had stars on it. And General Patton in it. He flew through the dust, up to our truck, and then past us toward the front of the column.

"You just about killed Patton," Duthie said slowly.

We packed the rest of the phosphorus grenades back in the case and sat quietly.

14

Nights in Castles

Charmes, France
September 11, 1944

We got word that Tex's four-year-old son had been killed in a car acci-
dent back in the States. He was devastated. "I'm over here putting up
with shit, getting bombed and shot at!" he cried. "*I'm* the one who's
supposed to die, not him . . . not that little boy."

Captain English said he was sorry, but that Tex couldn't go home.
We all felt terrible. The Red Cross was supposed to help servicemen
get home in case of emergency, but Captain said it wasn't possible
right now. I knew how Tex felt to some degree, because Grandpa Sacco
had died while I was in basic. But still, a little son who hasn't had a
chance to live his life is probably a lot different than a grandfather
who has.

We met more and more resistance as we moved closer to the city of
Charmes. The Germans didn't want to give up the town without a
fight. Unfortunately for them, fighting was precisely what we had
come here to do. After two days of unrestrained hell, the enemy
retreated, and we moved in.

Many civilians had been killed in the final conflict. The place was
destroyed and smoldering, but at least it was ours. Allied tanks rolled
through the cold, rainy town and blew up any possible enemy positions, while the infantry flushed out the remaining pockets of Nazis
hiding here and there.

Dead German soldiers were draped out of windows, blocking door-
ways, and littering the main street. Some had been there for days and
were starting to swell. A group of French soldiers came out of a build-

ing that had been something of an enemy headquarters, comparing trinkets, stuffing their pockets, and laughing. They walked up to a stack of German corpses in the street and set about looting them. The American MPs fired their rifles in the air.

"Get away from those damn bodies!"

The French looked defiantly at the Americans but then backed away, laughing among themselves.

The rain started coming down harder as GIs bearing the bodies of U.S. soldiers made their way into the intersection. They respectfully placed the dead side by side on the ground and stood watch over them until a series of trucks came and took them away.

Two days after our arrival in Charmes, Sarge came around to tell us that we were going to a show put on by the USO. We donned our Class A uniforms, folded neatly in our duffel bags since we'd landed at Normandy, and headed for a huge abandoned warehouse near an airstrip at the edge of town.

The place was packed. Guys were literally hanging from the rafters. There was a stage set up at one end, lights and flags everywhere, and a band playing lively show music as we entered. The opening act featured scantily clad girls—all legs and boobs—singing, dancing, and kicking high into the air in unison. The guys were going crazy. Some even got up and started dancing in their seats.

Then the girls parted, and through them walked Bob Hope. Like us, he was wearing a Class A uniform, except that he unbuttoned his jacket to reveal a bright orange polka-dot tie. The men howled with laughter, because the Army made such a big deal about being out of uniform. He did some jokes, then introduced "a friend of mine who I met on the road to the war." It was Bing Crosby. They sang a couple of songs and did some routines before introducing Dinah Shore. She sang and played around with them and a few of the boys in the audience.

"Hell, I know her!" yelled Averitt. "Dinah! Hey Dinah!" He was, of course, drowned out by all the other guys yipping and yelling at every move she made. "I'm telling you, I know her!" he insisted to those of us in his vicinity.

There was more joking, singing, and dancing, and then a big finale with streamers and confetti onstage and the showgirls running into the crowd to kiss some of the men.

"Come on!" Averitt said as he grabbed my collar. "Come on, guys! We're going backstage!"

"Full of shit, Averitt," Cardini said. "Just as full of shit as always."

"Hey, if you don't wanna come, then don't, smart-ass," he said to Cardini. "Come on, Sacco. Let's go say hello to Dinah."

He took off in the direction of the stage. I followed. So did the rest of the gang, including Cardini. Averitt was headed to the VIP area when an MP stopped him. "That's far enough, soldier," he said seriously.

"I'm a friend of Dinah Shore," Averitt said confidently. "I just want to say hello to her. She knows me."

"Yeah, and I'm a friend of Franklin Roosevelt, and you're not going anywhere." The MP put his body between Averitt and the stage.

"But she knows me," Averitt insisted. "Look . . . look, I'm from Nashville, and I'm a friend of hers from back home."

"Send her a note," Cardini suggested before turning to the MP. "He really does know her. She sends him letters all the time." Cardini loved the idea of bullshitting someone.

"Really?" the soldier asked. He studied Averitt's face for a moment. "Okay, give me your name, and I'll check."

He disappeared behind the curtains. Within a few minutes, he reappeared, and we heard a woman's voice behind him. "Paul Averitt!" It was Dinah Shore. "Well, Paul," she drawled. "How are you? Come over here and hug my neck!"

Paul moved in her direction. The rest of us followed as the MP stepped aside and let us pass. "Didn't believe me, huh?" gibed Cardini as he entered.

It turned out that Paul and Dinah had known each other's families back in Nashville. She was great to all of us, signing autographs, giving us hugs, and introducing us to Bob, Bing, and, most important, several of the showgirls.

Before we left, Bob Hope came over and said, "Hey, listen guys, I was talking to this GI the other day who told me he and a couple of friends of his were walking down the street in Paris with a few French beauties on their arms. Well, one of the soldiers cuts loose with this big, loud fart. His buddy says, 'Hey, it's not very polite to fart in front of the girls!' And he said, 'Ah, that's okay. These girls don't understand English!' "

We laughed like hell. Bing Crosby came up and said, "Well, that's not the best one. I heard from this soldier just out here that his grandfather was in this old folks' home back in the States. Yeah, you see, he was old, but he still felt young. So he was walking out in the gardens one day, and he goes up to another man and says, 'Hey, buddy, let's see if you can guess how old I am.' And the other fellow says, 'I don't know. About sixty-five?' And the old man says, 'I'm eighty-seven years old.' The other guy says, 'Hell, you look really young for your age. I would have never guessed that.' So the old man walks a little further, until he sees an old lady sitting on a bench. He walks up to her and says, 'Hey there, I bet you can't guess how old I am.' The old woman waves him forward until he's standing right in front of her. Then she reaches up, unzips his pants, sticks her hand inside, and starts feeling him up. She closes her eyes and keeps feeling around for a while. Then she pulls out her hand, looks the old guy in the face, and says, 'You're eighty-seven years old.' The old fellow's amazed. He looks down at his unzipped pants, he looks at the lady, and says, 'How'd you figure that out?' 'Oh,' she says, 'I heard you tell that guy back there.' "

We laughed until we pissed. Then Bob came back and told us another one. Then Bing. We were getting our own private show, not only from the two comedians but also from the showgirls who had gathered around in their revealing costumes to listen.

The troupe was leaving soon to put on another show in another town, so we said our good-byes, got our hugs from Dinah and a couple of the girls, and headed back to camp. Averitt, for his part, didn't seem quite as full of shit as he had before.

We moved out the next morning, deeper into the forest and hills north of Charmes, where we set up a bivouac. The German Luftwaffe was surprisingly bold, flying freely over our position, uncontested yet not attacking us. Then again, we were pretty well hidden among the trees. By midafternoon artillery shells were coming in, striking dangerously close to our camp.

"These things don't sound right," Cardini noted, referring to the distinctive whining noise the German bombs made when they were flying. About that time one exploded about three hundred yards from where we were sitting, sending dirt and rocks and tree parts high in the air.

Huge artillery guns, hidden in the area, began firing back. Assuming the infantry didn't know we were among them, Captain radioed HQ to get some clarification and instructions. The *incoming* shells, he was told, were being fired by the American artillery. And the *outgoing* shells were being fired by the Germans.

"Pack up your shit," Sarge said as he came running through the camp.

We quickly and quietly moved out along the same path we had taken in until we came across a U.S. infantry patrol. "Where the hell are you guys coming from?" one of the soldiers asked. "Nothing but enemy in that direction."

We had somehow managed to set up camp behind enemy lines without getting killed. Of course, we didn't know where we were, and, more important, the Germans didn't know where we were. And though ignorance can at times be bliss, it was good to be back on our side of the war.

Lunéville, France
September 20, 1944

Fighting was fierce as we moved out of Charmes toward the Moselle River, this time doing our best not to wander off behind enemy lines. The infantry had broken through while we were back in camp, and the engineers had thrown up a pair of bridges across the river to replace the ones the Germans had destroyed in their retreat. Sergeant Carapella, from HQ, told us that two men from Company B had been killed pulling cable across the river, one shot clean through the helmet and the other when a bullet struck a grenade on his belt.

We pulled within ten miles of Lunéville, a pretty little town in the hill country of eastern France. The Germans had surrounded it and were putting up a vicious fight in order to hold it. But we had too much firepower and were more than willing to unleash it. The infantry blasted through the enemy fortifications and gave the Nazis two options—either get the hell out of town or get killed. The ones who escaped disappeared into the mountains to join other units. We would catch up with them later. In the meantime we had a great deal of work to do, stringing wire all the way from Charmes to here and wherever else the infantry might be—which was everywhere—throughout the region.

In the first hours after the town was taken, we were walking down the main street toward the square when we heard the rat-a-tat-tat of bullets striking a stone wall to our right. White chips and dust flew up. We dropped in our tracks. Puffs of smoke were rising from the church belfry at the end of the street.

"Son of a bitch!" Chicago yelled.

"You hit?" Sarge asked.

"Nah," he answered, his head covered and in the dirt. "I don't think so. Don't the infantry know these assholes always hide in the damn bell towers? It seems like *somebody* would know that!"

As he was speaking, a small missile flew from a position in the square, smoke trailing behind it until it hit the opening at the top of the tower. It exploded with such force that it sent four Germans flying through the air. One side of the structure was blown away by the blast. Another corner crumpled down onto the street. The bell, still held aloft by a massive but fractured wooden beam, was ringing—not like a church bell normally rings, but more like it was in shock.

We got up and dusted ourselves off. "Sarge," Chicago said, continuing from where he'd left off, "everybody in the whole entire world knows there's gonna be snipers in a church tower, don't they? So why the hell didn't the infantry just blow the son of a bitch up before we came walking down the street like a bunch of assholes?"

"Speak for yourself," said Averitt.

"Not really the infantry's primary objective to go around blowing up churches," Sarge answered. "Not good PR."

"I'm just saying that somebody—"

"I think he *was* walking like an asshole," Cardini interrupted.

"Why don't they just put you in charge of the damn war?" Silverman asked Chicago.

"Yeah, that's a good idea. Put the asshole in charge, Sarge," Averitt agreed. "Then we can all go home and start teaching our kids German."

"I'll talk to Eisenhower," answered Sarge. "In the meantime just shut up and follow me."

Later that day we set up house in a castle on a hill above the town, generously sharing the expansive grounds with the infantry, seventeen tanks, and most of the XV Corps, including General Haislip and his staff. The complex included servants' quarters, a greenhouse, a stable, and even a small chapel. The main château had twenty or thirty rooms, the largest and ultimately the most popular of which had a fireplace. Large paintings and tapestries dominated the walls. Between them hung the trophied heads of deer and bears and a few animals I couldn't name. The place had been hastily evacuated by the Nazis, who left behind such items as cooking utensils, field glasses, and a few dozen bottles of good cognac.

Since the infantry and the tanks got there before us, we ended up sleeping in the garage. Some of the townspeople brought up pillows and blankets, which we welcomed and appreciated. After sleeping outside for so many months, we felt like we were living in luxury, especially since the garage was directly adjacent to the castle, almost touching it at one end. It was roomy, warm, and relatively comfortable.

It didn't take long to enter dreamland that night. We didn't realize until early the next morning that there was a large garden on the other side of the garage. That normally wouldn't have been a problem, except that in this case the garden was home to two extremely noisy mortar batteries.

They began blasting at first light, knocking us from the best sleep we'd had in ages, shaking dust from the walls of our quarters, and creating a curious rattling sound overhead. I opened my eyes to see the morning's blue sky above me. The mortars fired again at some target deep in the valley. The timbers shook, and the roof rattled.

"Tell me that's not glass up there," I said to whoever cared to listen.

"Yep," replied Averitt, lying on his back a few feet away, staring up. "Yep, it is."

The pillow felt too good. I shut my eyes and covered my face with my arm.

"Time to rise and shine, sleeping beauties of the 92nd Signal Battalion!" Sarge bellowed as he entered the garage. "Welcome back to the war!"

We spent the next two days going out into the hills in small patrols, setting up and testing phone lines for the lookouts, then returning to the castle at night. From the rooftop we could see enemy troops moving through the valleys. The mortars were firing around the clock at the Germans, so we stuffed cotton in our ears and hoped the roof wouldn't crash down on us.

Sergeant Carapella and his guys had set up their switchboards in the back end of the castle. Unlike Silverman and Cardini, who got most of their facts out of their asses, Carapella got his information directly from the horses' mouths. And he told us that we were about to have some very interesting guests.

"So I pick up the phone and it's this fella Sanders," he began. "And he says, 'Hickory, I got Etousa-2000 here, and he wants to talk to Lucky-6.' So I ring up Lucky-6, and the guy there answers and I say, 'Hey, I got Etousa-2000 here, he wants to talk to your boss.' The next thing I know, Ike is on the damn line—"

"Ike?"

"Ike," Carapella answered. "Eisenhower. That's his code name—Etousa-2000. And Lucky-6, that's Patton. So Ike, he gets on the line, and he says, 'George, you old son of a bitch,' and I says, 'General Eisenhower, General Patton's not on the line yet.' So Eisenhower just piddles around for a second—I can hear him breathing and writing something down—then he says to me, 'How are things at the front, soldier?' So I says, 'Great, General. Couldn't be better. Guys are doing a helluva job up here, sir.' By that time Patton picks up. 'Ike, you son of a bitch, how are you?' Hah! Ike was gonna say the same thing to Patton, but Patton beat him to the punch. Anyway, I hung on the line to make sure the connection was good, and I heard them set it up so they would come up here for a meeting. Them and Bradley and a few of the other big boys."

Sure enough, the place was swimming with generals within twenty-four hours. They met in the large room with the tapestries and the fireplace. I don't know the nature of the meeting or what they decided. I only know it lasted about four hours, during which time the infantry was guarding the castle and the mortar was pounding away at anything that moved in the distance.

When the meeting ended, Eisenhower, Patton, Bradley, and Haislip walked out on the side where a few of us were standing around with some infantrymen. We dropped our cigarettes, snapped to attention, and saluted. The generals gave a quick salute back, hopped into their waiting vehicles, and sped away. It was hard not to be a bit starstruck.

"You just saw some history walk by you," Silverman effused.

"I'm willing to bet this meeting was completely top secret," Sarge said, picking up his cigarette from the ground and wiping it off.

He was right. The meeting had been a secret—except for the ten thousand or so people Carapella had told.

The generals' cars were reaching the entrance of the village at the bottom of the hill when an incoming enemy shell struck the front lawn of the castle, digging a large crater and knocking a jeep and two trucks on their sides. We ran to take cover behind a retaining wall as another missile blew up one end of the stables. A third veered to the side of the château and exploded into the greenhouse with a tremendous crash, hurling lethal shards of glass high into the air and across the yard. It seemed like it took a minute or two for all the glass to fall to the ground, while we crouched closer to the wall and covered every vulnerable part of our bodies.

Our mortars, along with several large artillery guns set up around the grounds, started firing at the enemy with a renewed vigor. Things slowed down after a few hours. But then the Germans started shelling us again. They were shooting from somewhere in the hills, but it was difficult to pinpoint their location. Sarge said they were probably moving their big guns around after each attack, hiding among the trees, waiting for their next opportunity to strike.

Among the rare and welcomed luxuries of the castle was an indoor toilet, over the door of which Hodges had posted a sign appropriately classifying it as THE THRONE ROOM. The Throne Room was a place a man could go to be alone, a highly cherished place where he could think, contemplate, and take care of business. And because it shared a wall with the makeshift office of the U.S. Prisoner Interrogation Team, it was also a place he could hear German POWs being grilled.

When I heard some forceful voices speaking to one of the prisoners in English, I leaned my head back closer to the wall. They were asking him something about the locations of the artillery in the hills, but he responded to each question with either silence or his name and rank. Finally one of the frustrated interrogators said, "Look, you see that American soldier standing behind you? Yeah. That one. Unless you answer me right now, he's going to take you for a walk. And he's going to come back alone."

Duthie was knocking on the door, so I had to finish up and get out. "Listen to this," I whispered as I left the room. "This kraut bastard's about to get his balls cut off!"

I raced around the hallway up near the MPs to see if the American was really going to take the Nazi for a walk. But they never came out. One or two of the Germans must have eventually coughed up some useful information, because we heard that our guys found the locations of the enemy guns later that evening and destroyed them.

15

When Hell Freezes Over

We moved out of Lunéville by mid-November, going north into the Vosges Mountains toward Saarebourg in pursuit of the ever-moving front lines. The weather was turning colder, and the artilleries on both sides were still blasting away day and night with earthshaking power.

We were walking behind the truck along a riverbank near the town of Saverne, pulling cable off the spool, when a group of snipers opened fire. Cardini and Hutter grabbed the machine guns and sprayed the bushes. All got quiet. We slowly began moving out from behind the vehicles when a few more shots rang out.

"Sarge!" Silverman yelled. "Chandler's hit!"

Cardini must have spotted something, because he leveled his weapon at one area within the thicket and unloaded a blistering firestorm, instantly ripping through leaves, branches, and any human flesh that might have been hiding behind them.

When the smoke cleared, we moved cautiously toward the bushes. "Chandler?" Sarge called out. "You okay? Silverman, what happened?"

"I'm hit in the leg, Sarge," Chandler answered.

"How bad?"

"I dunno, Sarge. It hurts like hell, and it's bleeding."

"Hodges, Duthie, somebody get a medic up here for Chandler!"

Chicago was by now right up beside the bushes, cautiously poking them with his bayonet. "Come out," he said, "before I blow you to hell and back." Cardini moved to the other side, holding the machine gun.

Sarge moved up beside him. "Come on out," he ordered the enemy hiding within. "The war is over for you."

Averitt, down on the ground next to me, said, "Uh-huh. I'm sure some krauts are gonna just walk outta there and give the hell up."

There was a rustle behind the leaves. We aimed our rifles and prepared to fire. Sarge, Cardini, and Chicago took a slight, but quick, step back. I could hear muffled voices saying something in German, almost crying. Two young German soldiers appeared, each barely sixteen years old, hands atop their helmets, stumbling out of the thicket. Sarge and Hutter grabbed one of them. Averitt and I got up and jumped on the other. Spotted Bear, Duthie, and Tex moved behind them into the bushes and confiscated their weapons.

"Two . . . uh, three more back here, Sarge, all blown the hell to bits," Tex said. "Nice shooting, Cardini."

The first German had a bullet in his arm and was doing all the talking. I couldn't understand him, but from the tone of his voice, it sounded like he was asking us not to hurt him.

"This one's bleeding bad, Sarge," I said. Ours had dark blood oozing from his lower right abdomen and was having trouble walking. He looked to be in a great deal of pain.

"Okay, watch this one here," Sarge said to Hutter before turning his attention to the bleeding one. "Let's get him to a medic. Get HQ on the phone," he called out to Sam Martin. "Hodges, come give us a hand here. How's Chandler doing? Chandler, you okay?"

We got Chandler and the two Germans to the medics. Chandler wasn't thrilled about riding in the same jeep with the guys who shot him, but he made the trip without killing anyone. Fortunately, his wound was clean. The bullet had gone completely through his leg, striking no bones. He stayed in a field hospital for about a week and then rejoined us at the front. He hobbled a bit at first, but he had a Purple Heart coming his way.

Ingwiller, France
December 4, 1944

They were saying it was the worst winter Europe had seen in decades. Freezing winds blasted across open fields and howled angrily through the forest. Snow and ice covered the landscape and painted white the windward sides of trees, houses, trucks, and men. Though we were issued ponchos, heavy parkas, and wool socks, there was no true warmth to be had as we marched higher into the swirling snow of the mountains.

At night we sought any form of shelter we could find—an abandoned farmhouse, a stable, a cave, or a shack. Even a hollow in the ground or a well-placed log could keep the wind and ice off our faces long enough for us to catch a bit of sleep.

Sometimes the wind would abate and thick, heavy snowflakes would float down through the trees, carefully covering everything—and everyone—on the ground. In the rarest of times, when the shelling stopped for a minute or two, you could hear the frozen crystals softly landing on each other, piling higher and higher. The branches of the giant firs and pines, heavy laden and bending down, occasionally dropped big clumps of snow onto the earth below. If it weren't for the bitter cold and the terror of war, it would have been a beautiful scene.

But moments like these never lasted long. Within seconds the bombing and the gunfire would start anew. We would wipe the ice from our hands, shake the snow from our tools, and get back to work. The wind would howl again, sending snowflakes sideways, stinging our faces and making it feel ten times colder than it probably was.

The only time it wasn't snowing was when it was raining. And the rain, which seemed colder than the snow, somehow melted it into slush during the day, then froze over again at night.

The bitter cold, the treacherous roads, and the constant blizzards were making it difficult to get supplies up to the front lines. Guys were running out of everything, including the two most important tools for survival—food and ammunition. Mail got through one day, thanks to a courageous—perhaps crazy—lieutenant who used a precious tank of gas driving his jeep through the snow to bring us greetings from home. In some ways the letters were more appreciated than a hot dinner.

Among my mail was a letter from Uncle Paul, in which he bluntly stated, "Dear Joe, Something terrible has happened here." There was no further explanation—just his name at the bottom.

I took the letter to Sarge. "Is there any way I can find out what happened?"

Sarge examined the short, cryptic letter, turned it over, held it up to the light, and then handed it back to me. "Nope. Who the hell would write a letter like that anyhow?"

The Battle of the Bulge
Near the German Border
December 15, 1944

In the second week of December, the Germans launched a massive counteroffensive, flooding into an area from northeastern France to Luxembourg to Belgium, forming something of a protuberance, or westward bulge, in the Allied defenses. The war had been hard fought but successful up until this point. Now it was taking an ominous turn, as the crippling weather conspired with the desperate Nazis in an all-out attempt to break the backs of the advancing Allied forces. The Germans had been saving up for this move, gradually retreating while marshaling everything they had for one powerful and daring blast into the heart of our invasion.

Captain English told us we were all considered infantry from this point on and that everything else could wait. Sarge came around and issued us extra battle gear, including more cold-weather clothing, extra ammunition, morphine shots, and hand grenades.

"We don't know how to handle these things," I said, trying to give the grenades back to him.

"You had training. I was there," he responded. "Strap 'em to your belt. You'll probably need them."

"Maybe we trained one day or something," I said. "But I don't want these bombs hanging on me. That boy got shot in one, and it blew him up."

"That's true," Silverman began. "Sergeant Carapella said there was this boy who—"

"I don't give a shit what Carapella said," Sarge cut him off. "Strap these to your ass and start walking."

"Hell, Sarge," I complained, taking the grenades. "It's not like we're not loaded down enough already."

"You'll thank me later," he said, and moved me aside.

We were heading north with loaded packs on our backs and hand grenades on our belts, walking beside the trucks, forming a line on either side of the road, as the sleet, snow, and bitter cold pounded us. Every few minutes we would feel the concussion of a blast or hear machine-gun fire and jump facefirst into the freezing slush of ice and mud beneath our boots.

I was wearing insulated boots, wool socks, two pairs of long johns, a wool lining in my helmet, and gloves, and I had a T-shirt wrapped around my face. Despite it all I was still freezing as I struggled along, my bones aching, my hands stiff, my eyes numb. There was no sun or sky, just swirling snow, solid ice, air too cold to breathe, and bombs. Lots of bombs.

On December 17, we came across an infantry unit returning from a battle up ahead. They walked slowly and quietly between us. They were bandaged, bleeding, and broken. Some of our guys were bold enough to ask obvious things, like, "How you guys doing?" They would mostly just shake their heads and look down. "Fucking krauts," one mumbled.

They weren't as well outfitted as we for the cold weather. Some had regulation boots, basic-issue coats, and no gloves. Equipment had been lost or destroyed during combat, and fresh supplies, including heavy winter gear, had never reached their positions. How they hadn't frozen to death, I couldn't figure.

One young soldier was walking slowly past me, shivering, hugging himself to keep warm. "Hey, buddy," I said.

"Yeah," he answered, stopping reluctantly.

"Here," I said. I took the wool lining from my helmet and gave it to him. He looked at me almost disbelievingly but then reached out his stiff, cracked hand and took it.

"Thanks, buddy," he said, looking down at the lining, holding it over his hands to get them warm. "Thanks."

Duthie took off his helmet and did the same for another infantryman. Chicago, Silverman, Chandler, Averitt, Spotted Bear, on down the line—even Sarge did the same. "That feels good," my new friend said, still holding the cloth over his hands.

"Here," I said, looking up in Sarge's direction before pulling two hand grenades from my belt and strapping them to his.

His eyes widened. "Don't you need those?" he asked.

"Nah," I answered. "You take 'em. You want 'em?"

"Sure, but—"

"You take 'em then." I saw Sarge watching me, but he didn't say anything.

"Thanks, buddy," the young infantryman said, now putting the lining in his helmet and tightening the grenades onto his belt.

"Let's keep moving, men, and let these soldiers get to their camp," Sarge said.

Rumors started floating around that Germans had infiltrated American lines and were posing as GIs, dressing like us and talking like us and then, when they had the chance, giving away our positions or simply killing us. We didn't know if it were true or just somebody's fears run amok, but it seemed reasonable to be extra cautious. Captain told us that use of the code words was of paramount importance.

As we moved farther along into the combat zone, we saw the bodies of American soldiers covered with ice, their frozen blood spilled onto the snow. Some had huddled together in a futile attempt to keep from dying in the cold. Others looked peacefully asleep as living soldiers, unable to dig foxholes into the hardened earth, used their bodies for cover from which to fire at the enemy. Here, in the bulge, it was simple—kill or be killed.

Enemy tanks, artillery, and soldiers were everywhere, pouring into the region, arrogantly overrunning ground that, only a few days earlier, had been ours. The bulge was becoming a hemorrhage, and thousands of Americans were being slaughtered in the process.

The images of the frozen bodies on the battlefield—and the fear that I could be next—would not leave me. Not since Normandy had I felt so afraid and alone. "Hey," I said as I huddled against the wall of an icy trench with the rest of the guys on a bleak and bitter snowy night, all of us struggling to get maybe one minute of sleep. "Hey," I repeated.

"Yeah, what?" Averitt answered.

"You know that phrase 'When hell freezes over'?"

"Yeah. What about it?"

"Well," I said. "It did. And we're in it."

December 20, 1944

We were getting word that our attempts to stop the Germans were proving futile. They had overrun Luxembourg and were blitzing through Belgium, heading back to France. Our freezing, starving men were being outnumbered, outmaneuvered, and outgunned as supply and support routes were systematically shut down by the aggressive enemy and the unforgiving winter. Some Americans surrendered or resorted to hand-to-hand combat but were ultimately butchered by the well-equipped Germans. Even Patton's advances to liberate the region were said to be halted because of a lack of fuel and supplies.

What had looked like an Allied victory only weeks ago was now looking more and more like total defeat. The jubilant Germans intensified their onslaught as the swirling, assaulting weather wreaked havoc on our hopes and our lives.

Captain English received orders that we were to move northwest through the mountains toward Luxembourg. We were walking up a winding, narrow mountain path, following the convoy of trucks through a blizzard when one of the lookouts waved us to a stop. Sarge had us come up alongside the trucks and hunker down. A German Panther unit seemed to be tracking us from a distant ridge, the big guns of the tanks moving in our direction.

Suddenly, from somewhere below us, American artillery guns began firing, not at us and not at the Panthers but at some unseen target. The German tanks slowly turned their cannons and fired into the steep valley. The Americans returned fire to the now revealed enemy. In an instant, cannons and mortars were shooting from all parts of the mountains to all other parts. Snow was shaking off the trees as the earth quaked beneath. The trucks started up with a rumble, and we took off running as fast as we could. The adrenaline rush was tremendous. Everybody flew past the trucks, and some of the trucks even passed each other. We slowed down only after we had rounded enough turns so that the explosions seemed muffled and distant.

We were panting as we came to a walk, each exhale creating a visible fog, each inhale stinging as it went down. My lungs felt like they had frozen.

"Sacco," Silverman said. "Hey, Sacco. Joe."

"Yeah, I hear you," I answered.

"Hey"—he took a gasp of air—"we . . . we were surrounded back there, you think?"

"No shit," Averitt responded before I could.

"There were Germans crawling all over that pass," Duthie added. "How the hell we didn't get wiped out . . ."

"You think we should've stayed and fought?" Silverman asked, still struggling to breathe.

"Stayed and fought?" Cardini wheezed. "What we should have done is haul ass, which is what we did."

"You don't think we're gonna get in trouble for running, do you? After all, we're considered infantry now."

"Infantry, schminfantry," Cardini said. "Look. You see Captain English and that lieutenant and the Sarge up there?"

"Yeah."

"They were running like bats outta hell. We had to run like double hell just to keep up with those bastards. You wanna stay there and fight, you're welcome to, but me, I'm getting my ass outta that shooting gallery."

"Silverman," I said, reluctant to talk because it required more breathing. "You see that carbine you got there?"

"Yeah."

"Uh-huh. Yeah, well, it's got little bullets this long." I held up my index finger and thumb to indicate the size of one round. "Okay. Those Panthers—they got bullets about the size of Chandler's leg."

"The moral of that story," Cardini completed the thought, "is run like hell."

We had heard that American soldiers in Bastogne were completely surrounded by the enemy and had run out of food and ammunition. Carapella told us that the German commander had sent a courier into the town with a letter demanding the Americans surrender. The American general in charge sent back a response consisting of one word: "Nuts."

In defiance of the horrific conditions and unrelenting weather, Patton was moving north to liberate the besieged city. Determined to use every weapon at his disposal, he summoned a chaplain and ordered him to write a prayer to ask God to give us a break in the weather. The padre wrote the prayer, Patton read it, and the clouds parted.

The clearing skies were soon streaked by Allied fighters, bombers, and supply aircraft as tanks, artillery, and ordnance began rumbling overland toward the battlefields. The battle for the bulge was still raging, but I had seen neither the sun nor the blue of sky in months, and I felt like something big was in the making. It was as if an oppressive evil had descended over us, shrouding us with snow and ice and fog and rain, killing us and handing victory to our enemies. And now God had intervened, opening up the sky and giving us another chance to win.

16

Monique

Fénétrange, Alsace-Lorraine
Christmas Eve, 1944

The sunshine didn't last long, but it was enough for the Allies to get reinforcements to the front lines and give our boys a fighting chance. General Haislip sent orders for us to move south to help fill in the gap left by Patton's advancing infantry. In addition, we were happily reassigned back to our original duties of being a signal battalion.

After a difficult journey through snow, ice, and mud—during which a tank slid off the road and almost crushed Duthie and me—we arrived in Fénétrange, a picture-book village set amid forested hills in the Alsace-Lorraine region of France. The town was home to about three hundred people, almost all of whom spoke not only French but also German.

The entire 92nd—Companies A, B, C, and HQ—rendezvoused there. We took over a series of old wooden barracks long since abandoned by the French army and probably not even desired by the Nazis. It was dilapidated and creaky, with windows broken, doors that wouldn't quite shut, and floors that seemed as if they would cave in at any moment. A cold, wet wind was whipping through the large room like it owned the place. We plugged up as many holes as we could with blankets and socks and set about the task of getting warm.

Hutter, Hodges, and Bird Turd disappeared for a little while and returned with a small, scraggly, snowy Christmas tree in which we took great pride. We propped it up using some boards from the front steps and decorated it with packs of cigarettes, Hershey's candy wrappers, and utensils from our canteens. Hodges crafted a star out of six forks, a shoestring, and a few pieces of chewing gum.

Cardini had stashed his record player in the tool chest of the truck, so he pulled it out, wound it up, and put on some music. I looked around the room when I heard the first strains of "White Christmas." A lot of the guys stopped talking and hummed along. To us the music meant thinking of home. And thinking of home meant we weren't thinking of the war. So we hummed along, focused on the music, and enjoyed the sweet, albeit brief, make-believe journey it afforded us.

A little later, when he put on "I'll Be Home for Christmas," I started feeling down. Most all of us got real quiet and looked at the floor, at our hands, or out the broken windows. When it finished, Chandler said, "Hey, buddy, play that one again."

A couple of the mess sergeants had come up from Lunéville with a truckload of hot food so that we could have a Christmas dinner. A few of the guys from Company B helped them serve up turkeys, dressing, potatoes, gravy, beans, and lots of milk. And schnapps. We ate and drank until we couldn't eat or drink anymore. I didn't really like the schnapps too much, but I, like most of the guys, drank it like it was water. It was Christmas Eve, and we were far away from the homes we had known so long ago.

The wind blew cold that night, shifting the snow through the trees, pushing clouds under starlit skies, past the moon and beyond while bombs echoed in the distance. Inside our dwelling we lay wrapped in sleeping bags and blankets on the cold and drafty floor, rifles at our sides, gloves on our hands, caps on our heads. Our Christmas tree rustled ever so slightly as an unseen breeze crept in through the walls and chilled the room even more.

I was very sleepy, but I stayed awake for a little while and listened to the sounds of that night. It was Christmas Eve. And there was no peace on earth.

Christmas Day, 1944

There was plenty of work to be done on Christmas Day, at least for those of us who hadn't developed food poisoning from the previous evening's dinner. Several of the boys spent the night vomiting and ended up in the makeshift infirmary at one end of the barrack. Averitt concluded that the turkeys we were served had probably been left over from World War I. For whatever reason, it didn't bother me. I was more homesick than anything else.

We spent the day connecting and repairing switchboards, circuit-tester boards, amplifiers, cables, phones, and the like, getting everything back up to speed after our venture into the bulge. A cold rain started falling in the morning and continued until late afternoon, when it slackened to a drizzly fog. It wasn't much of a Christmas, but we were thankful to be alive.

When mail call came around a few days later, I was handed a letter from home in which Mama told me that Grandpa Amari had died. The letter was over a month old. This, I concluded, was the news Uncle Paul hadn't fully disclosed in his previous, mysterious note.

We heard that French troops were soon to arrive in the area, so we gladly offered them the barracks and went into town to look for lodging with the locals. It didn't take long to find a family willing to let a few of us set up residence in their cow barn, which turned out to be much warmer and more solidly constructed than our previous quarters.

Auguste Renard was a small, mustachioed man with a friendly smile and a welcoming attitude. His wife, Camille, was a gregarious and pleasantly plump woman who laughed easily and loved to cook. Both spoke a little English—just enough to make themselves understood. They had two children—a boy, Bernard, and a girl, Colette, both preteens.

As we were stowing our equipment in the barn's loft, I noticed that Chandler had gotten involved in a friendly conversation with our host family. I put my pack away and went over to investigate. Silverman, Averitt, and Duthie followed me. We were stunned to hear Chandler and the Renards chatting a mile a minute in German.

"Uh," Silverman said.

Chandler looked at us, then rattled off something to the French couple. They smiled and nodded to us.

"Chandler," Averitt said, getting his attention.

"Huh?"

Averitt lifted his hands and raised his shoulders slightly.

"Oh," Chandler said. "I was just telling them that you guys were my best buddies." He turned back to the couple and said something. They laughed. "I told them you don't speak German."

"Hold on," Duthie said. "You can speak German?"

"How in the hell can you talk German and you can't even read or write?" Silverman asked.

"Who told you that?" Chandler responded. I tried to look in another direction.

"Aw, hell, Chandler, everybody knows that," said Averitt. "It ain't a big deal. So answer the question. How do you know how to talk German all of a sudden?"

Chandler laughed. "I've been able to speak German since I was a little kid. I learned it from my grandmother. She was from Heidelberg."

"Ain't *that* a son of a bitch?" replied Averitt.

Chandler said something to the Renards, at which they cackled. Madame then invited us inside their warm house for coffee and some little French cakes. Chandler kept the conversation going the whole time. I really didn't understand most of what was being said, but I did enjoy the snacks. When we finished, the lady went into another room and came back with an armful of sheets and pillows to make our stay in the barn more comfortable. This, we could tell, was going to be a good place.

It was beginning to look like we would be here awhile. We didn't have much information regarding troop movements, but Sarge told us that we had a lot of work to do here and that we didn't need to worry about the rest of the war. "If General Eisenhower has an important conversation with anybody or makes any major decisions," he said, "I'm sure Carapella will let you know all about it."

It was deep into the night when I heard a yell. "Soldiers!" the thickly accented voice called up to the loft. "Help me!" It was Madame Renard. Her husband was rushing into the barn behind her. We scrambled down the ladder and gathered around. A cow was in the throes of the birthing process and was on her side, kicking, moaning, and in obvious pain. "Do not look!" the lady cried. "Help!"

Duthie and I were the closest to the animal, so we rolled up our sleeves, jumped down next to Monsieur Renard, and asked him what to do. He said something in German to Chandler, who in turn told Averitt, Silverman, Cardini, and Chicago to hold the cow's legs so that she wouldn't kick us. He then said to Duthie and me, "We must pull it out."

Apparently the birth was breech, and the calf was struggling to be born. "I was raised on a farm," I bragged to Duthie as we reached in with Renard, grabbed a leg or whatever we could grab, and began pulling. The tug-of-war took several exhausting minutes, during which time Madame Renard ran around looking from different angles, calling out status reports. Finally, and not without great effort, the calf, along with lots of other stuff, popped out of the cow.

I staggered away from the area, dizzy and sick to my stomach, and sat on the steps of the house. I had been there a couple of minutes when an extraordinarily pretty girl approached with a warm, wet towel. "Here," she said kindly in her accented English. "This is for you."

"Oh . . . thank you."

She gently placed the towel on my forehead before disappearing into the house. She quickly returned with a dry one and a tureen of warm water, which she set down in front of me. She motioned for me to wash my hands. Once I'd done so, she sat beside me, took the dry towel from her lap, and wiped my hands, holding them in hers to warm them. She took the wet, cooling cloth from my head and patted me with the dry one. "You feel better now?"

I felt better than I could ever remember feeling. She was smiling and was, without question, the most beautiful girl I had ever seen. She had dark hair and clear, fresh skin and the most sparkling blue eyes I could imagine. And when she smiled, dimples appeared in her cheeks, making her look more like an angel than a mere girl. "You feel better?" she asked again.

"Uh . . . yeah," I stammered. "Yeah, I feel good now."

"*Bon.*" She patted my hand, took the towels, and started to get up.

"But—" I said. She stopped. "But you don't have to go, do you?"

She smiled. "No," she said, sitting back down beside me. "Not yet. I will stay with you a little more."

She glanced at me, then looked shyly away. I found myself mesmerized by her face, so I made myself concentrate on her hands. They, too, were beautiful, each finger perfectly formed and perfectly feminine.

"You"—I pointed back at the house—"you live here?" I wondered why I hadn't seen her before now.

"No," she answered. "My family lives just there." She was pointing to the next house down the road. "We heard the noise with the cow, and we came to help."

"My name is Joe."

"*Enchantée,* Joe," she said sweetly, then pointed to herself. "Monique."

Monique. I had never heard a name so lovely in my life. We talked for a few minutes before her parents, who had been visiting with the Renards, came along. We said our good-byes, and she shook my hand. As they walked away, she looked back and waved twice.

I went to my bed above the cows that night feeling refreshed, excited, and, for the first time in forever, truly alive. I had met Monique, and I knew that I would never be able to get those sparkling eyes and that beautiful smile out of my mind.

The next day Sarge assigned me, Duthie, and Tex to string some communication lines from an observation post in the hills back to the Air Corps antiaircraft center in town. The job required their dropping me off and having me pull a spool of wire through the forest up to the lookout post, connect a phone, then meet the truck at a rendezvous point on the other side of the hill. Duthie was at a small switchboard we had rigged up on the back of the truck, calling me every few minutes to make sure I was okay.

We had set out in the late afternoon. Now, as I worked my way up and down hills through the thick woods, it was getting dark. True to form, it started raining. I pulled out my poncho and kept moving.

I hadn't heard from Duthie in a while, so I checked my map. It seemed like I should have found that observation post by now, at least according to my calculations. I was running out of wire, and it was getting late. The rain was really coming down now, that freezing, chill-you-to-the-bone type of rain. Each of the millions of icy drops was magnified in volume as they all crashed into countless leaves on their way to the ground. Cold little rivers were beginning to cascade down through the woods. I could barely hold the map out in front of me, let alone make sense of it. I crouched against a tree, got under the poncho, and lit up a cigarette.

After a few minutes, the phone rang. "Yeah," I answered.

"Sacco?" It was Duthie.

"Yeah."

"Where the hell are you?"

"Just sitting here having a smoke, trying to keep from drowning."

"You okay?"

"I'm okay. I'm kinda lost, though."

"Sacco, look, we— Tex dropped you in the wrong place. I think you might be on the wrong side."

"What?"

"You might be behind enemy lines."

"WHAT!?" I yelled, then caught myself. "What!?" I whispered.

"Just stay where you are. We'll go back to town and get some help."

"Bull*shit!*" I said. "I'm following this damn wire back outta here the way I came in."

I was up in a heartbeat, retracing my steps, moving quickly through the torrent, following the wire up, down, around, over, and out until I came to the road. A pair of headlights flashed up ahead.

I crouched down behind some bushes, peering through the darkness and pouring rain, trying to decide if the vehicle was one of ours or one of theirs. I inched closer until I could make out the star on the side of the truck and Duthie in the passenger's seat. The lights flashed again, and I made a dash for it.

Once I was on board, Tex turned the truck around so fast that it seemed like it must have lifted off the ground. I shivered all the way back to Fénétrange. When we returned, Sarge told me that a special shower truck had arrived there earlier in the evening and that I could go over and take a long, hot shower. Nothing ever felt so good.

The next morning Captain English called me into his office and apologized for the incident. I appreciated the apology but didn't think it was really necessary. In fact, the shower had felt so good that I was willing to do it again.

Madame Renard cooked a big dinner that weekend and invited us guys from the barn as well as Monique and her parents. She said it was to thank us for helping with the calf. I imagine the dinner was good, but I don't remember the food as much as I remember everything about Monique. She was dressed up and even prettier than I remembered her. She was sitting across from me at the long table but not directly across, about three seats down.

The other guys did their best to put on a show to impress her, but it wasn't working. All through dinner she kept glancing up at me, flashing me that smile, listening intently and translating whenever I said something, and making sure my plate was full. I couldn't keep my eyes off of her.

"You gonna take some cake or what, Sacco?" Cardini asked.

"Oh, thank you," I said to Madame. I hadn't noticed she was beside me with a small tray of pastries she'd baked especially for the occasion. Monique's father looked at me, then at his daughter. She smiled, picked up her fork, and took a bite of her dessert. She nodded to her father that the pastry was good. He looked back at me, winked, and smiled.

Her mother was Italian, from a town just north of Rome. Like Monique, she was slim and shapely, with a certain uncommon grace about her. She was quite beautiful for her age. She didn't speak English, but I was able to talk to her and her husband in Italian.

When the evening ended, Monique shook hands with all the guys before coming up to me. "If you have time," I began, summoning up my courage, "maybe we can go for a walk tomorrow afternoon."

Her eyes lit up. "*Oui.* Yes, Joe. I will walk with you tomorrow."

A couple of the guys were teasing me as we bedded down that night. They were probably just jealous—and they had a right to be—but I didn't want to hear anybody making jokes about Monique.

"Don't let 'em kid you, Sacco," advised Duthie. "A girl that pretty wouldn't give one of these monkeys a second look."

"If any of these fools gives you a hard time," Chandler added, "I'll shoot 'em." He pulled back the action on his rifle.

I snuggled under my blankets, hugged my carbine, and went merrily off to sleep.

I barely had time to return from my daily assignment and get cleaned up a bit when I saw Monique walking up the road in the direction of the barn. She was wearing a blue dress and a gray sweater and had one of those little thingamabobs in her hair on one side. I put on my jacket and went out to meet her.

She extended her hand as if to shake mine, and I took it. But then she walked closer to me and stood there, her hand in mine. She smelled as good and clean and fresh as she looked. "Monique," I said.

"Bonjour," she replied. There was a moment when nothing was happening. "Shall we walk?" she asked, her expressive blue eyes smiling.

"Yes, yes," I said, turning and releasing her hand. Her hand touched mine again, perhaps accidentally, as we began walking side by side.

"Comment ça va?" she asked. "How are you doing?"

"Oh, okay. Just spent the day putting up with lots of bullsh— Just . . . good. How about you?"

"Fine," she answered, slightly laughing.

We walked for about half an hour, talking sometimes in English, sometimes in Italian. She asked about America and what it was like. And I asked about what her country had been like before the Germans took over. She laughed at my jokes, touched my arm, and made me feel comfortable. I tried to look at her face whenever I could, but didn't want to stare. *I want my children to have blue eyes,* I was thinking when I noticed her rubbing her arms to keep warm.

"You're cold," I said.

"Yes," she said. "A little."

I took off my jacket and draped it around her shoulders. "But now *you* will be cold," she said.

"That's okay," I answered. "I'm used to being cold."

"Here," she said. "Give me your hand."

I reached out and took her hand. She stopped and turned slightly toward me. With my other hand, I gently touched her hair, then the side of her face. She smiled and looked down.

"Do you waltz?" she asked.

"Huh?"

"Waltz. Do you waltz? It's a dance."

"Yeah, I know. Uh . . . no. I've never done that dance."

She laughed. "My father has a phonograph, and he has a recording of 'The Blue Danube.' You know this music?"

"I think I've heard of it."

"Strauss. Johann Strauss."

"Yeah." I smiled. I wasn't sure what she was getting at, but I thought it wise to agree.

"He taught me to waltz," she continued. I must have looked puzzled because she clarified, "My father. And I can teach you . . . if you like."

"Yes," I responded. "Of course. I've been meaning to learn to waltz, but I've been kinda busy with the war and all."

"Good," she said. "It is settled. I will teach you. Look. Here is your first lesson." She faced me, got real close, then took my left hand and placed it around her waist. "Okay," she said, taking my right hand, "now we do this." She looked down and we began to move. *"Un, deux, trois, un, deux, trois, un, deux, trois . . ."*

I didn't know how to do the dance, but she was touching me, and it felt incredible. If this was waltzing, I wanted to waltz forever.

Later that night, back in the barn, I asked the guys, "Hey, somebody tell me—what exactly *is* the Danube?"

"Where are you from, bumpkin?" Cardini asked. "It's a river. It's real blue. You ever heard of that song, 'The Blue Danube'?"

"Yeah, of course I've heard of it," I answered defensively. "I was just asking."

Near Fénétrange
February 4, 1945

"Move this hunk of shit up a little, you big, ugly Texan asshole!" Chicago yelled as he banged on the side of the truck with his fist.

"Hey, how about watching your goddamned language?" Averitt said.

"I take offense at that," Spotted Bear added as they pulled the rest of the wire from the truck. "That big, ugly asshole's not from Texas."

I was halfway up a solitary telephone pole near the middle of a field, preparing to splice the lines together. "How you doing up there?" Chicago asked. "You ready?"

"Yep," I answered as I arrived at the top. I dropped one end of the rope onto which they would tie the line.

Chicago banged the side of the truck again. "That's enough. Hold up."

"That's something I never have understood," Silverman said to our driver. "They call you Tex, but you're from Montana."

"So what don't you understand?" Tex asked.

Silverman thought for a moment. "Nothing."

"You wanna know what I don't understand?" Chicago asked.

"No," Averitt responded.

"Okay, he's got it," Chicago informed Tex. "Let's take the rest back up to the road."

"What?" Averitt asked.

"What?" Chicago asked in return.

"You said you don't understand something. What is it?"

"So now you're interested in what I got to say?"

"Not really."

"Okay," Chicago said as he, Averitt, and Silverman jumped on the back of the truck. "Where the hell does Sarge get off telling us that we gotta do Company B's work?"

"It ain't Sarge's decision," Silverman began explaining as the truck pulled away.

"Like hell," Chicago growled.

"You need anything else up there?" Duthie asked me.

"Nah, I got it," I said as I began twisting the wires together. "Just take me a minute or two."

"Come on, let's go," he said, motioning to Spotted Bear. "We're going up to the road," he called out to me.

They were about twenty feet from the pole when I heard the unmistakable sound of German planes coming over the hills. Duthie and Spotted Bear started running, caught up with the truck, and jumped on the back as Tex sped toward some trees. Cardini, who had been waiting with Hutter in a jeep up on the road, grabbed the .50-caliber machine gun mounted on the back and started firing away at the planes. They, of course, started firing back.

I unbuckled my belt and slid to the ground as the planes swooped in my direction. They strafed on either side of the pole, then across the dirt toward the truck, which was by now disappearing into the woods. As soon as he penetrated the trees, Tex turned to the left and started driving serpentine. A few shots across the hood of the jeep had convinced Cardini to cease firing. Now he and Hutter had jumped from the vehicle and were running for cover behind a rock wall.

I, on the other hand, was standing completely stiff with my shoulder to the pole, trying to be invisible. The planes reached the edge of the field, turned, and started coming back for me. It was then that I heard a most familiar sound, one that would have been quite pleasant had I been hearing it under different circumstances. A large dog was barking, playfully bounding across the field in my direction.

I was trying to motion for the dog to go away, but he didn't seem to understand hand signals all that well. He got about four feet from me and launched himself through the air, almost knocking me over as I tried to stay aligned with the telephone pole. He panted and snorted and ran around, apparently wanting me to chase him. "Get the hell away!" I yelled, doing my best not to move. He was coming around again. And so were the planes.

They swooped down, spraying bullets across the field. I grabbed the dog, put him between my legs, and held on tight. If he were running around, he would have surely been killed. Several rounds hit the pole, causing it to reverberate. I shut my eyes as the dog yelped and tried to free himself from my grip.

Once the planes zoomed over, I released the dog and took off running. There was no way I could have made it to the trees, but there was a railroad track alongside the field near the pole. About seventy-five yards down the tracks was a tunnel. On the tracks was a wooden dolly. My accelerated thought processes told me I couldn't reach the tunnel in time using conventional means, but if I could get the dolly moving . . .

The planes were banking around, so I didn't have a lot of time to go over the plan. I reached the dolly and started pushing. There was a slight gradient down toward the tunnel, so it moved easily. The planes were straightening out. I pushed the dolly with everything I had, getting it up to top speed. The planes were coming. I could hear the bullets striking the field and charging in my direction. The dolly was rolling fast. I released it, then began chasing it. I heard the dog howl. With every ounce of power in me, I jumped.

There are moments in life when all you can say is, "Oh, shit!" This was quickly developing into one of those moments. As I flew up and through the air, it became rather apparent that I was sailing over the moving apparatus and was destined to land not on it but somewhere on the tracks in front of it.

I shut my eyes, reached out my arms, and prepared for impact. My head pounded into the wood of a railroad tie. My right hand landed on the track and was smashed as the dolly ran over it. Bullets ricocheted off the rocks lining the tracks. The dolly rolled on, impeded a bit by my hand but making its way into the tunnel.

I laid there on the tracks, my hand bloody and broken, my head a little confused. The Germans must have thought they'd killed me, because they flew away. Averitt, Duthie, and the rest of the guys were running toward me. I tried to get up so they could see I was okay.

I woke up in a hospital. A doctor was talking to a nurse. "Just give him a hundred cc's of—Well, hello young man," he said, turning his attention to me. "How do you feel? You're in the hospital. You'll be okay." He looked back at the nurse. "A hundred cc's of penicillin to stop any infection. Yeah, you broke some bones in your hand and suffered a bit of a concussion, but you should be in tip-top shape before too long. Anything hurting on you?" He poked his own stomach and side.

I shook my head no, so he poked my side. "Yeah," I said. "It hurt when you did that."

"Okay," he said, and wrote something down on a pad. "That's what I thought." He finished writing and hung the chart on the end of the bed. "It's nothing," he said. "Just a little shrapnel. Couple of little pieces of rock or something. We took it out. Might be sore for a day or two. You'll be good to go. I'm Dr. Bradley."

"Thanks, Doc," I said.

He began moving toward the little curtain that separated me from the other patients, but then he stopped, turned, and said, "Oh, yeah. You've got a visitor. Pretty girl. I'll send her in if you feel good enough."

She was glowing, smiling, and carrying a small vase of flowers. Never had I seen anyone so beautiful. She leaned over me and kissed my forehead. "I've been here to see you earlier, but you were still asleep," she said sweetly. "I brought you these to cheer up your room. I'll put them here in the window."

The flowers were nice, but it was seeing Monique that cheered me up. We'd known each other only a few weeks, but I'd changed a great deal in that time. The days had gone by quickly, yet the transformation within had been profound. Since landing at Normandy, I'd seen horrors on a magnitude I could have never imagined and had been changed by the sadness, the anger, and the fear that surrounded me. Doubt and uncertainty had become my constant companions. Sorrow and death had haunted me without reprieve. But now I felt different in a way I didn't know it was possible to ever feel. With a word, with a gesture, with a smile and a kiss, Monique had touched my heart and reconstructed my soul.

We spent hours talking about nothing, about everything, she sitting on my bed, me touching her hand, looking at her eyes, studying her profile and the way she breathed and how she would sometimes just tilt her head or look at me and smile. As the light from the window slowly dimmed, she leaned down, kissed me, and laid her head on my chest. "I am hurting you?" she asked.

"No," I responded as I touched her hair and breathed. "No, you're not hurting me."

"I must go now," she said. She leaned up a bit and pretended to wipe a tear. "I'm sad to go."

"Me, too," I said, before putting on some bravado. "Doc says I'll be up and outta here in no time, though."

"Joe?"

"Yes?"

She laid her head back down on me. "I was so scared when they told me you had been hurt. I was so scared." I hugged her and patted her back. "Please don't get hurt again," she said softly, a genuine tear falling onto my chest.

"I won't," I promised her. "I won't."

Doc came in early the next morning to check on me and give me some paperwork to fill out so that I could receive my Purple Heart. "It's pretty straightforward. Your injuries are listed as being received in battle. I'll be back around to collect it later this afternoon. Oh, yeah, here's a pencil."

He handed me the papers, took the chart from the end of the bed, and began writing something. "You're doing pretty well," he said. "I'm getting you out of here tomorrow morning."

"Really?"

"Yeah. They . . . I believe they . . ." he was checking something attached to the front of the chart. "Yeah, they have reassigned you to the 42nd Infantry. That's the Rainbow Division. Ass kickers. We'll make sure you get some R&R first. Okay, I'll be back later to see how you're doing and to get those papers. Take your time. Fill 'em out right."

Later that morning, from my bed, I spotted a jeep from the 92nd Signal Battalion pull up to the hospital. The soldier, a guy named Frank Serio, hopped out and ran into the building. When he returned with a bag of mail, I was in the jeep.

"Where the hell did you come from?" he asked.

"Through that window over there. Now, start driving."

He laughed and took off. "Are those flowers for me?"

"Very funny."

"Of course you know you're out of uniform."

"Yeah, I know," I said as we sped away from the hospital.

"Sacco, you're wearing a gown," Silverman observed as I entered the barn.

"What are you doing here?" Averitt asked. "Sarge told us to pack up your stuff."

"Where is it?"

"It's over there. They're gonna pick it up."

"I ain't going nowhere," I said as I began pulling out my uniform.

I was in Captain English's office within ten minutes. "What are you doing here?" he asked.

"I'm supposed to be with the 92nd, Captain," I answered.

"No, Private," he answered. "You've been transferred out to be with the . . ." He searched for some papers.

"Yes, sir, I know," I said. "With the 42nd Infantry."

"That's right."

"But I'm not going."

"How's that, soldier?"

"Look, Captain," I began. "I trained with the 92nd, I came over here with the 92nd, and I'm leaving with the 92nd." English was looking at me but wasn't speaking. "And if they try to send me to some other unit, I'm coming back here to be with the 92nd."

He studied my face for a moment. "Damn it!" he finally said. "Now I've got to do all that goddamned paperwork over again!"

I smiled. "Thank you, sir."

"Okay, get back to your unit," he ordered.

"Yes, sir. Thank you, sir," I said as I ran out.

Most of my life in Fénétrange revolved around Monique. Every minute not spent stringing wire, repairing cable, or testing equipment was spent with her. If all the training, if all the being away from home, if all the hardships of the war had been for the singular purpose of meeting her, then they had all been worth it.

And I wasn't the only one to find a girl there. Hodges had struck up a romance with the daughter of the mayor. She didn't speak any English, and he knew only what French he'd studied in high school, so they weren't able to have much in the way of meaningful conversations. Still, they liked each other's company enough to go for afternoon walks.

Their strolls were chaperoned by the girl's mother, little sister, and aunt, who followed several steps behind the lovebirds, watching their every move. Within days the escort grew to include an uncle, a couple of cousins, and the mayor himself. Eventually Averitt, Duthie, Hutter, Chandler, and other members of the 92nd joined the parade while Hodges and his mademoiselle happily walked along smiling but hardly saying a word.

"When they first met," I told Monique as we watched the entourage from across a small park, "Hodges tried to say, 'How do you do?' in French, but he got confused and said, 'How do you want to do it?' instead. She slapped him across the face, and they've been taking walks ever since."

"Perhaps that's why her family makes that big escort," she said, laughing. "Joe?"

"Yes?"

"Are you going away soon?"

"I don't know," I answered. "I don't suppose the Army will keep us here too long. This is the longest we've stayed anywhere so far."

"No," she said, putting her head on my shoulder. "I mean will you go to England for school?"

Silverman, Hutter, and I had been selected to return to England to attend Officer Candidate School to be trained as second lieutenants. "Oh, no, we're not going," I said. "I don't know about Hutter. He might go, but Silverman and I said no."

She picked up her head and looked surprised. "You can say no to your army?"

"Hell yeah, we can say no," I said. "They're using second lieutenants as point men in combat, and that means one of them is getting killed every four seconds. Captain said we could decline if we wanted to. So no thanks. I'm staying right here. I'd rather be a living private than a dead lieutenant."

She snuggled closer to me, and I put my arm around her. "What do you want to do after the war?" she asked, looking across the park as Hodges and his parade disappeared in the distance.

"I guess I want to go back to Alabama and see my parents and live there. I don't know, get married and have a family, I guess."

She settled even closer and looked up into the trees. "Alabama. What's it like there?"

"Well, it's real hot in the summer. Real nice, though. Real nice. I don't know . . . hills, kinda like this, trees, birds—just like this. In fact, it's *exactly* like this," I said, pointing around us.

"Oh, so there are trees and birds—like this?" she laughed.

"And hills," I corrected.

"And hills, of course. We cannot forget the hills, can we?"

"And the birds there, they all sing in English." I loved to make her smile just so I could see those dimples.

"Can I come there and see it someday?"

"Yeah, sure. You can come," I said. "You'd like it there."

"Then I will," she said as she kissed my cheek, "because I want to hear those birds singing in English. And if you are there, it will be a perfect place." She sat back and sighed. "Alabama."

There was, as always, a lot of work to be done. But, considering the circumstances, life was relatively good. The weather was warming a bit, and the days were getting sunnier. We were still billeted in the barn. I'm sure we all smelled like cows—or maybe the cows were starting to smell like us. Either way, nobody seemed to notice or care.

Me (second from left) and some boys from the 92nd taking a break in France, 1945.

I was getting pretty good at waltzing. Monique and I played that record of "The Blue Danube" at least ten thousand times. Her father also had recordings of other Strauss waltzes. They were nice, but "The Danube" was our favorite.

In addition, I was promoted to corporal, an offer that I did not refuse.

March 6, 1945

A group of us had spent most of the morning up in the hills repairing some lines. We finished earlier than expected and were walking back toward town when we came across a young boy from Fénétrange with a problem. His donkey, laden with supplies, was parked in the road and wouldn't budge. We decided to give him a hand.

"Come on, boy," I said to the donkey, pulling the rope. The animal didn't move. The more I pulled, the more he resisted.

"Give me that damn thing," Cardini said, taking the rope from me and pulling with all his might. There was no movement. "Hey, Averitt, Duthie, one of you!" Cardini yelled. "I'll pull up here, and somebody get behind it and push."

Averitt shook his head. "I'm not getting behind that damn thing. They got a nasty habit of kicking the shit out of things."

"I'll pass," said Duthie.

We pushed, pulled, threatened, pleaded, and begged, but the donkey didn't move.

"Let me kick that donkey's ass," said Spotted Bear, who then proceeded to run up and actually kick the donkey's ass. The donkey turned and looked at him as if he were crazy.

Finally, after every form of cajoling had failed, Sam Martin made a bold claim. "Are all of you ladies done? Because I guarantee you that I can get that damn thing running down the street in no time."

The rest of us stepped back and let him have his go at it. He went to the side of the road, pulled up some sticks and dried grass, came back, and placed them under the donkey.

"Actually," Averitt said, pointing to the donkey's head, "his mouth's over there."

"Not trying to feed him, Averitt," Sam Martin said. "Trying to make him run."

"And it's working," Cardini said.

"Yeah," Duthie chimed in. "I can hardly keep up with him."

"Patience, men, patience." And with that, Sam Martin pulled a book of matches from his pocket, struck one, and lit the grass on fire. The donkey glanced around at him but didn't think much of his scheme. As the flames rose, the animal fidgeted a bit. When they rose up almost high enough to lick his belly, he brayed, kicked, and took off galloping down the road with his young owner laughing and running behind.

March 12, 1945

I was climbing into the loft after spending the evening with Monique and her parents. The guys were involved in a somewhat-less-than-intellectual discussion.

"I remember during one of the short-arm inspections back in England," Silverman began, "the doctor told us about all the diseases you can pick up in those places."

"You're not listening to what I'm saying," Cardini said. "This one is approved by the Army."

"The Army approves whorehouses?" I asked.

"Up in Morhange," Averitt clarified.

"Line of guys about a mile long outside the door," added Chicago. "All the way down the block and around the corner."

"How the hell can the government approve a whorehouse?" I asked.

"I don't know, and I don't care," responded Chandler. "Just tell me where to get in line."

"I'm in," Spotted Bear said.

"Save me a spot," said Sam Martin. "But just let me go in front of Chandler. He looks messy."

Silverman shook his head, "Yeah, well you better wear a rubber, that's all."

"*You better wear a rubber,*" Chicago mocked. "*Or your dick will fall off.*"

"You'd have to *have* one before it could fall off," Averitt said.

"What about you, Sacco?" Chicago asked.

"What about me? I'm not interested. Have a good time."

"No, I mean, you been getting some action with that girl? Hey, if it was me—"

"Go to hell," I said.

"I'm just saying, if it was me, she'd be so knocked up by now that she couldn't see straight."

"How about if you shove it up your ass?"

"Are you telling me," he began, "that you have not poked that gorgeous girl in all the time you've spent with her?"

"That's enough," Chandler warned.

"Is that *all* you think about?" asked Averitt. "How about laying off?"

"Well, hell yeah, that's all I think about," Chicago replied. "What the hell else is there to think about?"

"There is this one little thing going on," Duthie said. "It's called the war."

"You may have heard about it," Averitt added. "It was in the paper."

"War?" Chicago mused. "There's a war? That must explain all those bombs going off and people shooting at me."

"I think Chicago's just here to get laid," I said.

"That explains a lot," said Silverman.

"I tell you what it don't explain," Chicago went on. "It don't explain how Sacco can get that doll when she could've chose a real man."

"Like who, for instance?" Duthie asked.

"Well," Chicago said, straightening his hair a bit, "like me, for one."

"Maybe she wasn't interested in going out with a horse's ass," replied Averitt. "That would explain it."

"Leave Sacco alone," Silverman said. "He saved my life a few days ago."

"Why'd you do that?" Chicago asked me. "What did he do, Jew boy?"

"He heard a Screaming Mimi incoming, and he called for me to get out from under this little stone bridge. And as soon as I cleared away from it—kapow! The whole damn thing got blown up!"

"What were you doing under the damn bridge?" Spotted Bear asked.

"I was trying to hide from enemy fire."

"Under a damn *bridge?*"

"Yeah."

"Hell," Cardini said. "You're dumber than I thought."

"What say we nominate Sacco for the Medal of Honor?" Averitt suggested.

"What say we kick his ass and tell him to mind his own business next time?" Chicago countered.

"We've all had a lot of close calls, haven't we?" Duthie reflected aloud, trying to bring some respectability back to the conversation. "We just have to pray to God to watch out after us."

"Hey, maybe Silverman can write a poem about it," Chicago derided.

"That's not funny," Silverman answered.

"Not as funny as that girly shit you write to your woman, I know that."

"You're not supposed to read other people's mail," Silverman fumed. "It's a federal offense!"

"Hey, look here, hebe. As long as I'm wearing this uniform and I got this here carbine, I *am* the federal government."

"You know what?" Chandler stood up. "Why don't I take you outside and whip the shit out of you right now, Uncle Sam?"

Spotted Bear jumped in. "Let me help!"

"Just calm down," Chicago laughed nervously. "I was only kidding the little Jew, that's all. No hard feelings, right, kike?"

Some of the guys—and particularly Chicago—had given Silverman a hard time after they found a letter he wrote to Rosanna. He let me read it, and it was poetry, but not the type that rhymes.

"Why the hell do you need me to help you write a letter?" I had asked him. "You're a damn good writer."

"Nah," he laughed. "I just like writing poems and things like that. But when it comes to writing down something that happened so that Rosanna or somebody can make sense out of it, I can't do that. Sometimes I can, but sometimes I get my words all jumbled up. That's why I like you to write some stuff for me. Plus, Rosanna likes to hear from you. Says you write funny stories. Then I can flower it up if I need to. It's sorta like when I make a piece of jewelry. You see, what happens is, I can draw it up and design it and all that, but then somebody will give me an idea and I'll incorporate it. Or maybe their idea's not that great, but I can fix it up and make a few changes here and there and come up with something good that somebody's gonna like when I'm finished. And then they put it on, or in this case read it, and they like it. See?"

"Oh, okay." I didn't understand what the hell he was talking about, but I thought it better to agree before he tried to explain it again.

March 17, 1945

Sarge came to the barn early Saturday morning and informed us that we would be pulling out of Fénétrange before noon, to get our gear packed up and on the trucks. The convoy was already starting to form in the street—engines revving, officers barking out orders, soldiers loading vehicles—as the push into Germany prepared to get under way.

It didn't take long to get packed and ready. Madame Renard brought us some fresh homemade bread for the trip, and Monsieur was helping us get our duffel bags down and into the truck. In the distance I could see Monique running from her house in the direction of the barn.

"Sarge told us to splice together four of the lines for the antiaircraft station and repair some wire before we leave," Duthie said. "We need to get on that."

"Tell him I'll be right there," I said.

He looked up and saw Monique coming. "Go ahead," he said. "I'll cover for you."

We spent the last minutes talking about how we would see each other again, about how I would come back for her after the war and how she would come to Alabama with me. She made me promise a hundred times that I wouldn't get hurt. I hugged her and tried to reassure her, but I didn't know what would happen to me as the war progressed from here. "I'll be back for you," I promised her.

We kissed, and she wept.

"Sacco," a voice said softly. It was Sarge, standing a few feet away. "It's time to go."

They had completed all the work without me. Now Monique and I embraced one final time as the truck waited patiently.

**Our convoy pulls
out of a small
town in France,
March 1945.**

I can never forget her face, her smile, her tears as we pulled away—
she standing beside the road, waving, hugging her mother's arm, looking directly into my eyes even as the distance between us increased,
wiping a stream of tears from her beautiful face and then . . . hidden
behind dust and trucks and men of war.

17

Like Crap Through a Goose

Zweibrücken, Germany
March 17, 1945

The artillery, the gunfire, and the earthshaking bombs all increased as we moved northwest and approached the defenses known as the Siegfried Line along the border of Germany. If the Germans had been unhappy about our taking over France, they had to be downright enraged by the notion that we were now invading their homeland. They put up quite a fight, but our guys unleashed a massive deluge of earsplitting, hellish fire on them, and they retreated. We were coming, and we were coming with fury.

By the time we crossed over into Zweibrücken, we were encountering relatively little resistance. But we had already seen this technique by the Germans in France. They would pull most of their troops out and leave a few pockets of snipers throughout the city to keep us busy. In the meantime the rest would circle and try to trap us inside. The artillery had planned on just such an eventuality. They bombed from the city limits outward as the infantry invaded the town. And the enemy fled.

The civilians here, unlike those in France, were not exactly happy to see us. In fact, they were terrified, apparently having been told that we would butcher them and rape their women. We certainly weren't here to harm civilians. In fact, we were here to protect them. But at the same time we were careful never to turn our backs on them. They considered us the enemy, and even the frailest of old men or the most unassuming of women would be more likely to hand us a knife in the heart than a bottle of wine and a flower.

It didn't really matter to us what they thought. We were just passing through, ready to move to the next town and kick some ass there.

The Nazis, unable to stop our advance, resorted to their old tricks
of blowing up bridges and cutting our wires at night, all of which
made for double duty as we stormed along. The infantry wasn't waiting for us. They were moving ahead, and we had to run to keep up.
We couldn't walk now as we had through most of France—there was
too much ground to cover. Instead we rode on the truck, jumped off,
did our work, jumped back on, and headed to the next town. We
moved quickly northeast through Kaiserslautern and Ludwigshafen,
scavenging existing phone lines when we could find them and stringing up new ones when we couldn't.

I always had a phone hanging from my belt when I was working so
that I could test the connection. I therefore decided it was worth a
shot to see if I could convince Sergeant Carapella back at the switchboard to help me rig up the line so that I could call Monique.

"My ass is grass if anybody finds out about this," he warned as I
pleaded with him from atop a pole. There was silence on the line.

"Hey, Sacco!" Averitt called out. "What the hell you doing up there?
We gotta make some tracks before sunset."

"Hold on!" I yelled. "I gotta test this line. There's a problem with
the—"

There was a ring on the other end. Then another. Then another.
"*Allô?*" It was Monique.

I covered my mouth close to the receiver and said her name.

"Joe? Joe? Is that you?"

We talked for only a minute, but it was a minute that brightened
my life. From high on that pole, as I listened to Monique's voice
through the crackling wire, the war beneath me seemed distant, and I
felt at peace.

"I gotta go," I said reluctantly. "Monique?"

"Yes?"

I covered the receiver and looked down to make sure Averitt wasn't
listening. "I love you."

"I love you, too, Joe," she said without hesitation.

The phone went dead. Carapella must have disconnected us. Probably had listened to the whole conversation. I didn't care. The most
important thing was that I'd said it. I had thought it over and over, but

when we pulled out of Fénétrange and I saw her standing there cry-ing, I knew I should have told her—I knew I had to tell her—that I loved her.

"What's taking so long up there, buddy?" Averitt asked.

"I'm coming down now. Hell, I had to make sure the line was working."

Worms, Germany
March 21, 1945

"This place is full of booze!" Cardini yelled into the phone. "Tell Tex to get the truck the hell over here!"

Averitt put down the receiver and got Tex's attention. "Cardini and Duthie found a shitload of booze. Let's get there on the double and load it up before somebody else finds it."

A split second later, we were moving, following the telephone line to a cellar on the outskirts of Worms. Inside, there were thousands of bottles of champagne, stacked neatly and covered with a fine layer of dust.

"Start loading," Tex said. "We're gonna need more trucks."

We loaded sixteen trucks full of the bubbly. There was already a lot of telephone equipment, and we had to allow some space for the men, but we managed to squeeze a few hundred bottles on each truck. We proceeded to celebrate our windfall by drinking it and passing it out to any infantrymen we saw on the roads.

The fun ended as we descended the hills into the city. Worms was one of the last towns before the Rhine River, and the Germans were not going to give it up without a fight. There was heavy bombardment as our artillery moved into position in the hills. Once they started bombing, the enemy quieted down, and the infantry began the attack.

We were already down in the valley waiting for orders to enter the city when the infantry came through. For lack of a better plan, we went in alongside. Fighting was heavy even though most of the place had already been blown to hell. German soldiers were still everywhere, fighting to the death, shooting it out on the streets and from the rooftops, lobbing hand grenades on our positions, blast-ing us with tanks, pounding us with everything they could muster as we worked our way through the town.

We got pinned down by some snipers about a block or two away from what was left of the city's main square. Sarge, Cardini, and Chicago fired some cover long enough for us to break out a window and get into a small storage room just below street level. Battles were going on all over town. From where we were, I could hear gunfire erupting up and down the street, overwhelmed repeatedly by the massive explosions of bombs.

Chandler had gotten trapped across the street, near the truck. The snipers still had a pretty good angle on his position and fired on him whenever he made the slightest move. Cardini had the idea of going back out and getting to a position from which he could give Chandler some cover. Sarge went with him.

They climbed up through the window, moved to the right, and barricaded themselves behind some debris. They then sent a hail of gunfire up at the snipers, who shot back with even greater ferocity. Chandler still couldn't move without drawing deadly attention. The battle went on several minutes before Sarge and Cardini jumped back through the window.

"Panther coming," Sarge panted.

A German Panther tank was going through an intersection a block away. Sarge and Cardini didn't wait around to see if it made the turn.

The gunfire from the sniper's nest stopped. Chandler looked around, then got up. "Get down!" Sarge yelled through the window. Chandler looked up at the window and began to move toward the front of the truck. "GET DOWN!" Sarge yelled again, waving his arms at Chandler.

Chandler hesitated and then sprinted out into the street toward us. "NO!" Sarge screamed.

A shell exploded a few feet away from Chandler. Dirt, rocks, and pieces of cement broke windows in the surrounding buildings and flew into the room in which we were huddled. Chandler was down, mortally wounded, a gaping hole in his abdomen, his left leg missing, blood oozing from his side and pooling onto the dirt below. I could see he was crying as he gasped for air. I felt my hands shaking.

Chicago started crawling out the window toward him, but Sarge pulled him back. "Sarge?!" he said, pointing helplessly out into the street.

Chandler was choking as he struggled to breathe. And then, with an effort almost superhuman, he began to move, pulling himself with his arms. Suddenly a hail of gunfire from the direction of the tank exploded on his body.

The street began moaning under the grinding weight of the Panther. Sarge told us to move up against the wall under the windows and take cover. I was hiding behind some rubble, next to a brick column. There was no more room beneath the windows, so Sarge motioned for me to duck down and stay where I was. A dim beam of sunlight traced itself across my chest. From my position I could see some of the activity in the street. The remaining panes of glass rattled as the tank approached.

Foot soldiers were running alongside it, looking for anyone who may have been hiding. A grenade from another building missed the Panther and exploded into the ground, sending shrapnel flying through the street. The tank stopped, turned its cannon, and fired. Most of the windows crashed down as the shock waves reverberated through the area.

The Nazis had hit the ground. Now they were back up, prowling beside the massive vehicle as it slowly began to move again. They were yelling in German, getting closer. I could see the front of the Panther come into view through the window. It blocked out the shaft of sunlight and shook dust from the roof.

The room was completely sealed off in the back by huge wooden beams, chunks of masonry, and other fallen debris, so we couldn't get out any way other than the way we'd come in. I reached into my pocket and pulled out the crucifix.

A young German soldier broke away from his crouched position alongside the tank and ran up to the windows. He knocked out a few of the remaining shards of glass with his rifle. Most of it fell onto Sarge and the guys as they held tight their positions along the wall. I pulled back around the column and tried not to breathe. All he had to do was peer inside long enough for his eyes to adjust, and he would see us. I held the cross to my chest and shut my eyes.

Amid the confusion on the street, another soldier began yelling something in German, barking out orders to the one near the window. The young Nazi turned away and ran back to the Panther, where he crouched down and continued to escort it as it rumbled along. I could see the shadows of ten or twelve more Nazis move across my chest as they chased behind the tank.

As the German vehicle entered the next intersection, an American Sherman tank, on the alert and poised a block away, blasted it broadside. The Panther jolted and shrieked, then tried to turn its guns toward the attacker. The Sherman blasted it again, then again, then again.

It was hard to tell what was going on from where we were, but there must have been more than one Panther in or around the square. Within twenty seconds U.S. fighter planes were flying overhead, bombing the German tanks and the area of the square into oblivion. Our building began shaking so hard that it seemed it would collapse.

Finally the bombardment ceased. I could still hear some small-arms fire in the area, but nothing like before. Sarge led us out of the basement, through the window, and back into the street. American infantrymen were already out patrolling, cleaning up any remaining pockets of the enemy who still wished to fight, gathering up any white-flagged ones who had thought better of it and decided it was time to surrender.

We left the town of Worms in rubble and ashes. For us the victory was muted and hollow. We had lost a friend. I had lost a brother.

I wrote a final letter and gave it to Sarge to send to Chandler's mother. And late that night, when we were on the truck moving out, when I was alone and had time to think, I wept.

Crossing the Rhine
March 24, 1945

The Germans were bombing their own bridges in a desperate attempt to keep us from crossing the Rhine. But the Army Corps of Engineers was putting up new ones as fast as the old ones could be destroyed. Nothing was going to stop us now. We were moving fast and picking up steam. We were going through the enemy, as General Patton had requested, like crap through a goose.

The Air Corps bombed Mannheim and Bensheim in anticipation of our arrival. We turned north through the countryside, putting up lines as fast as we could, making sure they worked, and then moving on to

The convoy on the autobahn near Mannheim, Germany, March 1945.

the next site. Armored vehicles crowded the road, kicking up dust and blasting any- and everybody who dared get in our way. As we moved eastward beyond Darmstadt, we encountered a surprising sight. Hundreds of German soldiers were marching toward us carrying white flags, trying to surrender to each vehicle in the convoy as they passed.

"Send 'em to the back," the American soldiers were calling out down the line.

"Don't worry," Cardini mumbled. "We don't have time to mess around with these bastards." He leaned over the side of the truck and yelled, "Keep moving, you sons of bitches!"

We bivouacked in a hilly and heavily forested area about thirty miles east of Darmstadt. Cardini, Averitt, Duthie, Silverman, and I were walking over to the mess tent when we saw one of the Red Ball trucks rolling into the area with a German soldier sitting on the hood.

"What the hell is this?" Cardini asked.

We went over to investigate. Sarge was already on the scene.

"I don't know *who* the hell he is," explained the soldier at the wheel. "Kept trying to give me his name and serial number and all that, but I kept telling him that I don't care. Said he wanted to surrender and tried getting in my truck, but I told him, 'Hell no, you ain't getting in my truck.' And so I told him, 'You can ride up there on the hood so's I can keep an eye on you while I drive.' So here he is. I'm dropping him off with you."

"You're taking him back," Sarge told the driver. Then he turned to the German. "Get back on the truck. Get right back up there."

"But where's I'm gonna take him, Sergeant?" asked the driver.

"I don't care. Take him anywhere. Take him back to headquarters. We don't have time for him here."

"But he's a pain in the ass, Sergeant! Keeps saying he's cold or something in Nazi talk. And he don't understand a damn thing in English."

"I'm sure he *is* a pain in the ass, Private," Sarge responded. "But he's *your* pain in the ass, not mine."

Aschaffenburg, Germany
March 27, 1945

We were on the back of the truck, stopping and going, rumbling slowly along in convoy, a jeep right behind us, a Sherman tank behind that, infantry soldiers—many of them fresh replacements just out of basic training in the States—walking alongside, trying not to choke on the dust.

(From left) Duthie, Paul Ciaccio, and Averitt in Germany, 1945.

"Why the hell are they sending such young kids up here to the front?" I asked Duthie. "These boys don't know what the hell they're doing. They're gonna get hurt."

"Whataya mean 'kids'?" he asked.

"Look at 'em!" They didn't have that tired, desolate look you see in the eyes of combat veterans, although some of them did look scared. Their equipment was new, and many of their faces had not yet seen a razor.

"Look at yourself, Sacco," he said. "You're only twenty!"

My God, I thought. *He's right.* I felt like an old man.

We could hear gunfire and bombing as we approached the edge of the relatively small town of Aschaffenburg. German troops were putting up a resistance. That wouldn't have been all that unusual had it not been for the fact that a vast majority of the local townspeople were fighting even more vociferously than the soldiers—wielding pitchforks, dropping burning bundles of hay, throwing knives, rocks, bottles, furniture, and anything else they could get their hands on, including homemade bombs.

Three stops on the way through Germany, 1945.

General Haislip ordered us to do something almost inconceivable at this point in the war—retreat. We pulled a safe distance out of town while a small patrol of American armored vehicles drove through the streets blaring messages to the townspeople that they should immediately evacuate if they wanted to live.

Just after sunset the artillery began a heavy bombardment. Silverman, Duthie, and I went around and lay down on the ground just in front of the huge artillery guns and watched them fire over us. It was quite a show until a captain in charge of the gunmen came around to ask us what we were doing.

"Just watching these things fire, that's all," Silverman answered.

"Get the hell outta here!" the officer bellowed. "What the hell's wrong with you? If a dud comes outta that thing, it'll kill all three of you. Get the hell back to your unit!"

We got up and ran away.

The next day U.S. planes dropped bombs into the heart of the town. The artillery shelled again that night.

The next morning after that, we walked into what was left of Aschaffenburg with no resistance, impeded only by the mounds of debris in the streets and the smoke rising from the rubble.

Though the adult German citizenry wasn't exactly enthralled with having American soldiers in their country, the children warmed up to us rather easily. Not only were they unaffected by the politics of their parents, they were also completely fascinated by the jumbo Hershey bars we carried in our backpacks. The chocolates were meant to give us quick energy on the battlefield, especially when rations weren't readily available, but they ended up serving as exceptional public-relations tools as children flocked around us to get a piece.

We did our best to look out for these little ones. Whenever we would see small children watching us in the chow line, we would each get a kid, put him or her in front of us, and give the child half of our mess kit. Captain told us we weren't supposed to do that, because there might not be enough food for all the military, but we didn't care. We loved these kids. They had gone through hell and really had nothing to do with the war. A lot of the children we encountered were orphans. We had plenty of food, so we would make sure that when they left, they didn't leave hungry. It made all the guys feel good to see these kids eat, to do something kind and gentle, in contrast to the chaotic inhumanity surrounding us. I know it made me happy, and I believe it made God happy, me watching over the children while He watched over me.

April 4–11, 1945

The towns became a blur as we battled furiously north- and east-ward, the Air Corps bombing a path for us through what still stood of Germany and its army. The Nazis continued to give us hell, fighting with all the deadly force remaining in their arsenal, hurling every bomb, every shell, every bullet in our direction in an increasingly desperate attempt to stop our advance.

We had no intentions of stopping or even slowing down. We would summon whatever firepower we needed, beat them into submission, do our jobs, and move on. The toll, though, was growing. Six men from Company B were killed in an ambush while stringing telephone lines outside a place called Lohr. And as we turned southeast and moved deeper into Bavaria, we heard more stories of German atrocities against American POWs committed in defiance of the Geneva Convention.

"Holy shit!" Cardini said as we walked through the streets of a little town named Bad Brückenau.

"What?"

"Take a look around!" he said. "Look at these damn men!"

Most of the older German men in town had tiny mustaches and hair slicked to one side. It looked like a village of Hitlers. They were going about their business, giving us suspicious looks, talking to each other in hushed tones.

"This is spooky," Silverman said. I noticed all our guys holding their rifles a little more at the ready as we walked along.

A group of us were walking through another small town when we spotted a dead German soldier lying on the side of the road, a hole through his head. Chicago used his rifle to flip him over. The soldier was actually a young boy, maybe twelve years old, in a Nazi uniform. The ammunition strapped to his belt consisted of wooden bullets.

There was a quick movement in a row of bushes a few feet away. Another boy soldier jumped out of the foliage and began running. None of us made a move to stop him.

"Take one bottle of cognac each and get some sleep," Sarge said as we returned from duty at the front lines. "Only one, Spotted Bear. And make it last a few days."

I grabbed mine, sat down against a tree, opened it, and took a sip. Work had been difficult. The fast pace of our advance was stretching us to our physical and emotional limits. I also thought a lot about Chandler, vacillating between grief, anger, and an impending sense of doom. We were winning the war, but the Germans were proving that a retreating army could be a very dangerous army.

I tried to patch through to Monique a few times while stringing phone lines but had no luck. Carapella told me there must have been some problem with the lines back in France.

I pulled out some paper and began writing a letter.

"You know what I was thinking?" Silverman asked.

"What's that?"

He stammered a bit, then said, "It's just that . . . well . . . we've seen lots of guys die—our guys and their guys."

"Yeah, I know," I said, putting the letter down and taking a swig of my cognac.

"What I mean is that . . . well . . . sometimes we've been right there and watched."

"What are we gonna do? We call for the medics. That's all we can do."

"That's not what I mean," he said. "I mean, when they die, what do they say?"

I thought for a moment. "Hell, I don't know. 'I don't want to die'? Something like that?"

"No," he responded. "And I've noticed. They all call for their mamas—in whatever language they talk. They always call for their mamas."

"Yeah," I said. He was right.

"We're all here fighting for our countries and our governments and all that, but nobody ever calls the name of the president or the Führer when they're dying. Ever notice that?"

"Yeah, come to think of it."

"They call for their mamas."

I took a couple more gulps of cognac, picked up my paper again, and resumed writing.

I finished the bottle before I finished the letter, and I was finding it increasingly difficult to breathe. In an attempt to locate air, I grasped the lowest branch of the tree, pulled myself up, and started climbing.

"Where you going, Sacco?" Spotted Bear laughed as he and the other guys watched.

When I got as high as I could go, I started singing. The guys on the ground were laughing and applauding, but Sarge came along and said, "If you fall out of that tree and kill yourself, I'm gonna have your ass court-martialed."

"Ah, Sarge," I yelled down, "you wanna hear an Italian song?"

Sarge turned to Cardini. "You got your carbine?"

"Yeah, Sarge, right here."

"Shoot him down," he deadpanned.

I stopped singing and felt around for the limb below. "Comin' down, Sarge. I'm comin' down now. Don't shoot."

"Hold your fire," he said to Cardini.

Somewhere during my descent, I—or perhaps everybody else—got turned upside down. I hung there for a while, trying to figure out what was wrong, until Spotted Bear and Duthie climbed to my rescue.

Bamberg, Germany
April 12, 1945

The German populace was not happy with the way the war was going and made every effort to let that displeasure be known. We didn't have anything against the civilians, but they sure didn't like us. No matter where we happened to be, we would always try to be friendly to the people, smiling, saying hello, and the like. They, in response, would either give us dirty looks or ignore us altogether.

After a while we just gave up. "Screw 'em," suggested Chicago. "They don't wanna smile? I say screw 'em. We'll do the smiling."

We really smiled when we got to the outskirts of Bamberg, where we found a railroad boxcar filled with hams, bread, cheese, butter, coffee, and other items, including cigars and lighters. We loaded up our backpacks, put the rest on the trucks, and carried it away.

A few miles later we came upon a deserted house that some of the guys were saying had belonged to a Nazi officer. A few infantry soldiers were milling about on the premises.

"Come on," I said. "Let's take a look at this."

The place was crammed with nice furniture and lots of odds and ends. There was a library full of German books and pictures all over the walls. This guy must have left in a hurry, because he didn't take anything with him. Even his uniforms were still hanging neatly in his closet and his shiny boots still lined up against one wall.

Guys were taking things they liked. I found a couple of real nice swords and an ivory-handled dagger, each with the swastika and the SS emblem on them. I tucked them in my backpack for safekeeping. As we were leaving, I walked into the parlor and spotted a full-size, ornate accordion.

"You can probably play that thing, can't you?" Sam Martin asked.

"Sure he can," Cardini answered. "He's a good Italian boy."

Between the loot from the train and the swords, I was loaded down with stuff. "Ah, what the hell?" I said, picking up the instrument and slinging it over my shoulder. "I'll learn to play it."

Silverman was sitting on the steps outside. "Where's the truck?" I asked him.

"They went on up ahead. Why?"

"Because I wanted to unload some of this shit."

"Oh. Where'd you get that thing?" he asked, pointing to the accordion.

I looked around at the house. "In there. Where do you think I got it?"

He shrugged. "Just asking. Hey, Sarge went up ahead with the rest of the guys to find a decent building or someplace for us to set up quarters. Says he's gonna park the trucks right next door to the church and for us to go on and meet up with them there. It ain't as far as it looks. Only about a mile and a half."

We were walking through the streets toward the city center, weighed down with our spoils, when I felt the wind of a bullet as it flew inches past my head. I instinctively fell to the ground facefirst. The accordion made a loud groaning sound as it landed heavily on top of me.

Duthie, Averitt, Silverman, Cardini, Spotted Bear, and Sam Martin turned their weapons in the sniper's direction and began firing. Chicago, who had been carrying a machine gun over his shoulder, swung it around and emptied it at the shooter. Pieces of glass and chips of stone were flying. From my position it looked like they were going to cut the building in half with bullets.

When the smoke and dust cleared, we could see that a woman had fallen to the street from her perch on the second-story ledge. She had been shot with several dozen rounds. Her sniper's rifle, itself shot and mangled, was twisted under her.

As far as we could figure, the woman had gotten off two shots before being killed. The first had narrowly missed my head, and the second had mortally wounded the accordion.

The night before we left Bamberg, a group of us were walking down a rather deserted street when we spotted an attractive lady in her mid-twenties walking toward us. She, unlike any of the other women we'd seen since the war began, was wearing makeup—lots of makeup. She approached without smiling and spoke. "You Americans, you are winning this war," she said in an ominously thick German accent. "But you won't win the next one."

"Yeah, and I got an ass you can kiss, too," blurted out Chicago.

We laughed. She didn't think it was funny. She looked seriously at Chicago, then down at the ground as if we were in her way. We made a space for her, and she walked through. We were still laughing when Chicago turned to take another look, presumably at her ass. He stopped in his tracks.

Averitt, Duthie, and I turned to see what was going on. "What the . . . ?" Averitt asked.

What we saw was . . . nothing. She couldn't have gotten more than two or three steps away but was nowhere to be seen. Along our side of the street, and stretching for several blocks, was an eight-foot wall. Across the street was an empty park. Within a few feet of us, she had vanished, disappeared without a trace.

We decided to call it a night and get back to our company.

Erlangen, Germany
April 18, 1945

Erlangen was supposedly home to a number of German military hospitals. Our mission was to capture the city and take the patients as prisoners, under which scenario they would have received better medical treatment than they were getting. Considering the reports of massive surrender among the German ranks, it was not outside the realm of possibility that they would have given up the city without much coaxing.

Unfortunately, the Nazi SS troops decided to fight to the death.

As we approached the city, we could hear a great battle going on. Sarge told us to fall back so that the infantry unit behind us could catch up with the one in front of us. Before we knew what was going on, we were in the middle of the fighting. I was in a foxhole with Averitt, Duthie, Spotted Bear, and a few guys I didn't know, shooting, covering up, shooting some more, and trying not to go deaf when the cannons from the American tanks fired over our heads.

The guys were spread out a bit. I saw a medic run to tend to Chicago, who was apparently hit by shrapnel. After a few minutes, the medic scurried to another foxhole, and Chicago resumed fighting. I really couldn't see at what or whom we were shooting, but I knew they were out there, so I leveled my rifle and blasted away.

An incoming mortar exploded a few feet from my left ear. In an instant everything went quiet. I could still see the other guys firing their weapons, but I could hear nothing. Then a painfully loud ringing started to reverberate through my head, and I began to feel very confused.

I stood up, resisting Averitt's attempts to yank me back down, turned, and wandered off, apparently babbling something about needing to climb a tree and string some cable for Sarge. I felt a rock hit me in the back. I turned. Duthie was frantically waving at me. He looked like he was yelling something, but I couldn't hear a voice, so I waved back, turned, and continued strolling along. Out of the corner of my right eye, I spotted Silverman coming at me, moving low and fast, then tackling me to the ground.

"Hey, Silverman," I said, somewhat dazed. My hearing was coming back, and the battlefield was loud. "You okay?"

"What are you doing, buddy?" he asked. "You decided you don't wanna live anymore?"

"No," I said, my head aching, my eyes barely able to focus. "I wanna live. I wanna live. But Sarge said we needed to—" I stopped and looked up into the sky.

"Holy shit!" Silverman yelled. "It's the cavalry!"

Allied planes were streaking overhead, swooping in on the Nazis, strafing, bombing, and doing all those nice things warplanes do. They continued until they had blown Erlangen, its SS troops, and its hospitals to hell.

My head was spinning, and I felt a bit disoriented for a few hours. As we rode along on the truck, Silverman kept holding up his hand in front of my face and asking, "Okay, how many fingers do I have up?"

"Look, I'm fine," I told him. "And if you do that again, I'm gonna shove all your fingers—*and* your thumb—up your ass."

He laughed. "Okay, quit playing around. How many fingers?"

Nuremberg, Germany
April 20, 1945

What we found when we entered the city of Nuremberg was complete and total devastation. Structures large and small were reduced to piles of rocks, maybe with one jagged wall standing here or there, windows shattered and staring blankly into open space. Smoke was slowly rising from the smoldering, bombed-out wreckage of buildings and vehicles, blocking the sun and sending everything slightly out of focus. A few civilians milled about, some calling out names as they searched for loved ones, some quietly crying, some silent. Cats poked and darted through the debris, trying to find food or a place to hide.

Bulldozers were pushing mountains of rubble aside as we slowly made our way through the streets. "What the hell was this like?" Duthie asked, looking around. We knew what he meant. The terror that must have rained down on the citizens of this place was almost unimaginable.

We stayed there one afternoon, just long enough to put up a communications board, and then moved out, south toward Munich. We were walking along the edge of an open field, smoking and talking with some infantrymen, when we heard a blast. Tex, walking a little farther out in the open, had hit a land mine and was flying through the air. In the space of the instant it took for us to turn and look, a volley of machine-gun fire erupted from a group of trees across the field. I could hear the thud of bullets striking nearby and felt blood splatter across my face.

Everyone dropped to the ground and returned fire. The infantrymen splayed out to fight the enemy. Amid the confusion and yelling, I noticed several men down and hurt. Cardini was clutching his leg, and Sarge, himself bleeding from the arm, was on the ground, slowly moving in his direction. Silverman, who had been walking beside me, was on his back, three holes torn across his chest, a fourth in his shoulder, blood gushing out, a terrified look in his eyes. "Sacco!" he gasped, reaching for me.

"Duthie!" I yelled. "Help!"

Duthie, only a few feet away, crawled to us. "Oh, shit!" he yelled. "Sarge! Medic! We need a damn medic right here right now!"

We pulled Silverman into the woods behind us, around to the other side of a tree. Duthie grabbed a cloth or T-shirt or something from his backpack and began pressing it to Silverman's chest. In an instant it was soaked red. He picked it up and looked at it for a moment, then put it back on the wounds and held it there. "MEDIC!" he yelled over the gunfire.

I was up against the tree, sitting, holding Silverman while Duthie worked on his chest. "Come on, buddy," I kept saying.

"We're gonna make it, Joe," he struggled to say.

"We're gonna make it, don't you worry," I said, trying not to panic. "We're both gonna make it! You said so yourself."

He looked at Duthie, still calling for the medic, who was now running in our direction. He turned his eyes back to me and mumbled something I couldn't understand.

"What?" I said, bending down, putting my ear closer.

He tried again, but he was just making sounds, not words.

"Save your energy," I said. "You can tell me later."

He reached his left hand up toward the sleeve of my jacket but then put it back down to his side. I held his head in my arms and tried to make him more comfortable. The medic had arrived and was giving him morphine shots. Blood was still pouring from his wounds.

"We're both gonna make it, buddy," I said. "Just hang on."

The medic, working frantically up until now, slowly looked up at me. Duthie, crouched against another tree, had buried his face in his hands.

"Hang on, Silverman," I repeated. "We're both gonna make it. We're gonna make it! You said so yourself! Come on, buddy." I was crying, rocking back and forth, as the medic quietly crawled away to tend to other people.

I pulled out the letter a couple of nights later—the one he'd written for Rosanna. I had promised him that I would send it, along with the wedding rings and the other jewelry, in case anything happened to him. I made up a small package, including a short letter from me to tell her how much I thought of Silverman, addressed it, bound it with string, and gave it to Sarge.

**The Danube River
April 27, 1945**

Tex was hurt pretty badly. They sent him back to a hospital in France, where they had to amputate part of his leg. Word was that it would take some time but that he would be okay and on his way home soon. The medics had been able to patch up everybody else well enough for us to continue on. Sarge's arm was in a sling, but he was still able to give orders with the other one. They gave Cardini the option of going to a field hospital, but, for reasons unknown, he decided to stay with the rest of us. The medics had extracted a bullet from his leg. It had apparently ricocheted off something and was going relatively slowly when it hit him, so it didn't do much damage. He put it in his pocket for good luck.

We had to keep working, setting up telephone stations and stringing cable as we moved along. There was little or no resistance until we got near the town of Donauwörth on the Danube River, where a few Germans who still wanted to fight shot mortars at us. An American armored unit traveling with us, some twenty-five Sherman tanks strong, blasted the hell out of the enemy from every conceivable angle. And all got quiet.

**Me near Bamberg, Germany,
April 1945.**

Later, when we stopped long enough to take a rare lunch break, I walked down to the edge of the river. The water wasn't beautiful and blue the way Monique and I had envisioned when we'd danced so many times. In fact, it was murky and green. I shut my eyes and tried to remember the way I'd imagined it to be. I missed Monique so much, especially right now. I knew she would be heartbroken when she found out what had happened with Silverman and Chandler. I looked at the river again, took off my helmet, and sat down on the bank, trying—but not quite able—to concentrate enough to hear the music in my head.

"Tough times," I heard Sarge say. I hadn't seen him walk up. He, like me, was now staring out at the water as it gently flowed past. He handed me a cigarette.

"Yeah," I answered, shaking my head and taking the smoke. "How's your arm there, Sarge?"

"It's okay," he said. "It's fine. I'm a tough old cuss. How are you doing?"

I put my head down and stared at the ground. "I'll be okay," I said. "Sarge, has there been a problem with us getting the mail from France?"

"I don't think so." There was a silent moment, and then he said, "You never know out here. Things get held up, rerouted, lost. Hell, a buddy of mine sent me a postcard from Paris, and it took a damn month and a half to get to Fénétrange. He could have walked it there quicker than that." He took off his helmet and sat beside me. "Here, need a light?"

"Yeah, thanks," I said. "You know, Sarge, one time we were in Féné-trange and we had gone out on a work detail. And Duthie and I were pulling some cable, and we were getting ready to haul it up in some trees, and this jeep pulled up, and it was General Haislip."

"Haislip?"

"Yeah, and he pointed to me and said, 'Soldier, where the hell's your helmet?' and I said, 'It's right there on the back of the truck, General,' and he said, 'Why isn't it on your head?' And I said, 'Well, General, because it's heavier than hell, and if somebody shoots me in the head, then I figure the helmet ain't gonna stop the bullet anyway.' "

"You said that to General Haislip?"

"Yeah, and get this: He said, 'You get your ass over there and put that damn helmet on your head right now. My job is to keep you alive. I don't need any of my soldiers dying because they won't wear the proper headgear.' And I said, 'Yes, sir.' And I don't think I've taken it off since then, until just now."

Sarge chuckled. I looked back down at the ground, then out at the water.

"Tough times," Sarge said.

"Yep," I answered, fighting back tears.

Rain, Germany
April 28, 1945

Beyond the Danube we started seeing German soldiers surrendering en masse, white flags waving as they approached us on the roads. Captain English said to just keep sending them back and let the armored divisions pick them up.

We stopped in the early evening and set up a bivouac north of a town called Dachau. It seemed a little early to turn in, but we were all pretty tired and looked forward to the extra rest. Before we hit the hay, Captain called us together, pulled out a piece of paper from a notebook, and said, "These orders just came in." He cleared his throat and began reading aloud. "'Tomorrow the notorious concentration camp of Dachau will be in our zone of action. When captured, nothing is to be disturbed. International commissions will move in to investigate conditions when fighting ceases.'"

As we headed back to our tents, Duthie said, "Must be some kind of prison or something."

All I knew for sure was that I was dead tired. And so, as the lights and colors left the sky on this final day before the liberation of Dachau, all of us—quite unaware of the gravity of the morrow's mission—settled into dreams of war's end and going home.

18

Where the Birds Never Sing

Dachau, Germany
April 29, 1945

There were small patches of snow still on the ground in this part of Bavaria, persistent reminders of the chilling winter that had dominated Europe. Low, gray clouds had rolled in during the night and formed a colorless blanket across the sky by the time we awoke.

We moved swiftly and uneventfully southeast, covering the remaining miles in a matter of minutes. The town of Dachau was small, quiet, and rather unassuming, with the exception of a large prison complex on its outskirts. This, according to Captain English, was our objective. And by the time we approached it, some infantrymen were already climbing over the walls.

The place was big. There was a brick wall, probably eight to ten feet high and topped with barbed wire, surrounding the facility. Spaced out along the wall every half block or so were guard towers, most of which looked abandoned. Beyond the wall I could make out the tops of other buildings.

I could hear gunfire as the infantry engaged the enemy within the perimeter. Cardini was riding in a jeep with Sarge, and some of the other guys were following behind. "Come on," I said to Duthie. "Let's see what the hell's going on."

We ran up to the wall and crouched down beside some infantry soldiers. "You believe these sons of bitches still wanna fight?" one of them said, checking the clip on his rifle, then motioning for the others to follow him along the wall to his left. The soldiers got up and ran to the end, then disappeared around a corner.

The gun battle didn't last long. After about twenty minutes, we heard American voices calling the all clear.

"Well, hell," Duthie said. "Is that it? That was easy."

"Yeah, I guess so," I said, getting up and motioning for Averitt, Spotted Bear, Chicago, and the others down along the wall to follow us. "They're opening the gates!" I yelled.

"How many krauts you think they took prisoner in there?" Duthie asked as we began walking.

"I don't know," I answered. "Big place. Bunch of 'em probably holed up in there like rats. The more the better. Assholes. I'd like to get my damn hands on a couple of 'em."

The other guys had caught up to us as we rounded the corner near the main gate of the camp.

"What the hell's that smell?" Averitt asked.

"That's what Nazis smell like when they're scared shitless," Chicago said.

There were five infantrymen quietly standing near the entrance. "You boys kicked some major ass here in a hurry, didn't you?" Spotted Bear commented. They didn't smile or laugh or even seem happy. They just looked down.

"Hey, what the hell is this place?" Averitt asked. "A prison or something?"

One of the infantrymen shook his head. Another looked up at us and said, "Welcome to hell."

We fell silent as we walked into the camp itself. The smell that had caught our attention as we approached was now overwhelming. Within a few steps all of us came to a stop and stared around in horror. No one said a word. I felt nauseated, dizzy, and confused. My brain couldn't comprehend what my eyes were seeing.

Everywhere I looked, in every direction, I saw the dead—women, children, old men, babies—beaten, starved, stabbed, shot, butchered, and left to rot on the ground. Most were wearing the tattered striped uniform of a prisoner. Others were completely naked. Some were so emaciated that I couldn't tell if they were male or female.

Many had been running when killed, falling into grotesque, contorted poses along the path. Some had been bludgeoned, their brittle bones broken and sticking through their skin. Others were decapitated or sliced open. One young boy—he couldn't have been more than ten or twelve—had been shot in the forehead. Even in death his brown eyes were opened wide, as if he had been staring defiantly at his murderers. A few feet away, a small group of children looked as if they'd been kneeling and then shot in the back of the head. A little girl clutched a piece of twisted cloth between her hands, obviously aware that she was next in the executioner's order. Her eyes were half open now, staring at the cold ground. A slightly older girl lay dead beside her, reaching her hand out as if to offer comfort.

Spotted Bear grabbed my shoulder and pulled me up off my knee in front of the children. "Look at this," he said, pointing to a long wooden ramp about four feet off the ground along the inner wall. Down its entire length were the dead bodies of those used as moving targets for the Nazis, who had shot at them with high-powered rifles from marked positions across the yard. Most of the victims were young, probably teenagers who could run fast and therefore made better sport for their killers. Many had fallen off and were piled onto the dirt in front of the ramp. A huge mound of bodies, taller than a man, leaned against the wall near one end.

Spotted Bear was breathing hard, his eyes darting back and forth across the carnage laid out before us, his face registering the shock and disbelief we all felt. Sam Martin walked up. "They captured some of the SS," he said. "They got 'em over in the other yard."

"Duthie, where're you going?" I asked as he passed behind me. Duthie pointed to some railroad cars parked just inside the camp. I turned and followed him.

American soldiers were standing near the tracks, peering into the boxcars. "They're all dead," one infantryman said to us as we approached. He hung his head in disbelief and walked away.

I climbed up on one of the open cattle cars and looked down into it. There, scattered on the floor below, were dozens of twisted, lifeless bodies. They looked like they had been starved to death—emaciated, mouths agape, eyes staring vacantly at the sky. We were too late for these people, and it broke my heart.

One soldier about seven or eight cars down yelled that he thought he saw someone move. A surge of adrenaline rushed through the area and several others ran in his direction. He jumped down into the car and started sifting through the bodies. Some of the other guys called

I looked into railcars at Dachau . . .

out that they were going to climb down and see if they could find anyone still alive.

I could see that everybody in this one was dead, so I maneuvered over to the next one and peered down. Duthie was climbing up to help me look. Inside, among the confusion of wasted and decomposing corpses, I spotted a woman, probably in her early twenties, leaning against the side, nursing an infant. Both mother and baby were dead—she struggling to give her very last drop of life to her newborn and he struggling to live, to enter a world that would not have him. I didn't see, I couldn't see, anything else. The Madonna and Child, the supreme symbols of life and hope, of all that is and can be good in our world, lay murdered beneath me. I knew that no matter how long I lived, I would never see anything so profoundly tragic as this.

I felt my hands trembling. I didn't know if Duthie or anybody was talking to me or if there was any noise whatsoever. All I knew was the lady and her baby. And somewhere in those endless moments as I stared down, transfixed on the scene below, I realized why we had left our homes and come to this distant land to fight a war. I just wished we could have gotten here in time. I just wished we could have saved that lady and her infant son. *May God forgive us for not getting here sooner*, I prayed.

. . . and this is what I saw. Germany, April 29, 1945.

I climbed down and stumbled away from the train, stepping over more bodies until I reached a small platform. I sat, took off my helmet, put my head in my hands, and, against my wishes, started crying. My hands were shaking so hard I could barely hold them to my eyes. Duthie had climbed down, too. Now he came, sat beside me, and put his arm around me. "It's okay, Joe," he consoled, his voice cracking and barely audible. "It's okay." I lifted my hand and pointed back toward the railcar. "It's okay," he said. "Bastards."

I wasn't alone. Many of the other men were crying. Some were enraged, angry, belligerent, stunned. Others were silently walking around with dazed looks on their faces as the overwhelming stench of death shrouded the camp. I saw a two-star general leaning against a wall throwing up. Even after everything we'd seen and experienced since landing at Normandy, this was beyond any hell we could comprehend.

Averitt, Hutter, Chicago, and a couple of other guys came walking by. "We're going into the POW compound," one of them said.

"Come on," Duthie said. "Let's go. That's where they have the prisoners."

I wiped my eyes, got up, and followed along. I didn't care where we were going. I just wanted to get away from the trains. Averitt had found a truckload of film a few days before, so he was snapping pictures of what we saw. I watched as he took a shot. He looked up at me. "They'll be historical one day," he said solemnly. "Got a camera?"

"Yeah," I answered. He tossed me a couple rolls of film.

Fashioned atop the wrought-iron gate leading into the prisoner compound was a sign reading ARBEIT MACHT FREI. We didn't need anyone to tell us what it meant: "Work Makes Free." "Look at this shit," Chicago said as he hit it with the butt of his rifle, and we walked through.

Inside were row after row of squalid, dilapidated barracks, out of which were slowly and reluctantly emerging fearful prisoners. Several thousand inmates were already crowded in a vast fenced-in area in front of the barracks. Beyond them, just outside their containment area but well within their sight, was a gallows from which were hanging ten or twelve prisoners. American soldiers had climbed up and were cutting down the bodies.

The inmates had apparently not understood what was happening during the first thirty minutes or so after we took over the camp. Now they were beginning to realize that we were the good guys and they had been liberated. Some of them began yelling and shouting. The cheers of the few soon grew to become a deafening roar throughout the camp. *These are the ones we save,* I thought as chills ran through my body.

They were crowding together, reaching through the fences, wanting to touch us. They looked like the walking dead, like they were too frail to stand or reach or even smile. Yet in their parched voices we heard jubilation, and in their sunken eyes we saw the joyful glimmers of resurrected hope. They were laughing, crying, singing, holding their arms high in the air, embracing each other while looking at us in disbelief, almost unable to grasp that their day of liberation, their exodus, was at hand.

Prisoners in the confinement area of Dachau, April 29, 1945.

Just then, as if on some celestial cue, the sun broke through the clouds, sending shafts of light and color over the entire camp. Darkness had been defeated by light. And the prisoners, squinting into that brilliance, reached out to us and wept tears of joy. Their agony was over. But for us, the liberators, our journey into this deepest part of hell was just beginning.

Averitt wanted to go to where the Americans were holding the captured Nazi SS troops. Most of the regular German soldiers had fled before we took over the camp. The SS, though, had remained, vowing to fight to the death and kill as many of the inmates as they could in the process. Our guys had them lined up in front of a wall just opposite a shallow stream that ran along the inside edge of the compound. As we approached, we could see the SS sneering defiantly at their captors, hurling insults and arrogantly laughing. An infantryman who had been watching—and whom Spotted Bear and Sam Martin identified as an American Indian—flew into a rage. He picked up a machine gun and began mowing down the Nazis. A couple of other infantrymen took the opportunity to pump some lead at the SS as three or four Americans subdued the Indian and took the gun from him.

Most of the Germans were dead. Some had collapsed against the wall. Others had turned and fallen into the stream, faceup, eyes open, cold and clear water peacefully flowing over them.

One of the younger ones was on the run, having dodged the bullets and exploited the confusion well enough to flee. He was now sprinting toward the edge of the compound, racing away as if he might really escape. In an instant, Averitt, Duthie, Chicago, Spotted Bear, Sam Martin, and I were in full stride behind him. He ran across a small bridge at one end of the camp and disappeared to the left, behind a lone building. Barriers and barbed wire prevented his going any farther.

We split up and captured him cowering down against the back wall. He looked like he was in his early twenties. "Let's kill his ass right here!" screamed Chicago, his eyes glaring, the veins in his neck bulging. He jammed the end of his carbine into the Nazi's neck. "You son of a bitch!" he yelled. "You bastard!"

The Nazi was crying and saying something in German, but he stopped when Spotted Bear slapped him across the face with his elbow.

Executed Nazi SS troopers in Dachau, Germany, April 29, 1945.

"Let's take him back," Duthie advised.

"I say let's kill him!" Chicago yelled again.

"We need to take him back to—" Averitt began.

"They're just gonna kill him anyway, so let's just do it here," Chicago said. I agreed. I wanted to kill him on the spot.

"Let's go, kraut bastard," Spotted Bear said, clutching the German's arm and jerking him off the ground and away from the wall.

Chicago reluctantly withdrew his weapon from the soldier's neck. "What the hell is this?" he said, taking a step backward to size up the building. It was one story tall and made of brown brick, long and narrow, with chimneys spaced out along its length. As we rounded the corner with our prisoner, we could see stacks of dead bodies being carted to the facility while smoke poured from the chimneys.

We brought the Nazi back over the bridge and were heading in the direction of the infantry when we heard a commotion coming from the prisoners within the compound. They had run up to the fence and were waving their arms, screaming, calling for us to bring this Nazi to them. The SS trooper glared at them with hatred and contempt.

"What?" an agitated Chicago shouted, jumping in the Nazi's face and walking backward. "You don't like these people? You don't like these people?! You should get on your fucking knees and beg for their forgiveness!"

The soldier spit at Chicago.

The prisoners were yelling, pleading for us to give him over. Two American infantrymen were guarding the gate a few feet away. "Bring his ass here!" one of them said. The emotion of the moment overpowered us. We pushed the Nazi toward the fence, the guards swung open the gate, and we shoved him in.

He spun and fell to the ground. Before he could get up, the male prisoners swarmed him, kicking, hitting, and cursing. One came up from behind carrying a piece of wood and began to pummel the Nazi, who tried desperately and futilely to protect his head with his arms. Once he lost consciousness, the man with the stick backed off, and the others, now joined by women and children, went into a frenzy, each landing blow after blow, each kicking, crying, and exacting their revenge. The bedlam continued for several minutes, until all had their requital. Finally some of the men came forward with heavy metal rods the infantrymen had given them and made sure he was dead.

We stood and watched the spectacle without uttering a word. As I witnessed the fury unleashed within, I found myself feeling a curious and powerful mixture of anger, exhilaration, curiosity, and even guilt. I had thought it would feel good to see this bastard killed, to see the prisoners tear him apart, especially after what he and the other Nazis had done to them. But seeing more death did not bring me the pleasure I had assumed it would.

As he was dying—as his avengers were beating him to death—I had become fixated on papers falling from his pockets onto the ground below. They seemed, perhaps in contrast to the furious whirlwind surrounding them, to be falling in slow motion . . . down, ever so gently onto the dirt of the prison yard. I wondered what evil information might be written on them.

When the men were satisfied he was dead, they backed away, and a few of the older women came forward and spit on the corpse.

He was alone, on the ground, surrounded by his papers. I reached down into my pack, pulled out my camera, and snapped a picture. The two infantrymen opened the gate, went inside, and dragged the body out. I stood quietly, staring at the papers left behind.

"He was gonna die anyway," I heard Averitt say.

"Huh?"

"The infantry was gonna kill him anyways," he said. "Might as well let *them* kill him. They deserve to. Son of a bitch had it coming."

"Yeah, I guess," I said. "Hey . . . those papers . . ."

"Come on, Sacco," he said. "Let's go."

The Americans had found a few more SS hiding around the place and were lining them up against a wall in another part of the compound

Nazi SS trooper beaten to death by prisoners in Dachau, Germany, April 29, 1945.

for execution. Averitt said he heard one of the brash Germans yell out, "You are required to adhere to the Geneva Convention and give us a trial!"

An infantryman yelled back, "Here's your trial, you bastard! You're guilty!" And with that he shot the Nazi in the forehead.

For understandable security reasons, the prisoners were required to stay within the confinement area. A special group of soldiers, doctors, and nurses was assigned to give them food, drink, and emergency medical treatment as needed. Some were said to be so emaciated that they died from trying to eat or drink too much at once. Others, weakened beyond help, died within hours of their liberation.

Some of the relatively healthy prisoners—those who had been here only a very short period before our arrival—were recruited to help gather the multitude of bodies from around the camp. They took baggage carts from the train depot and set about the grim task of collecting the dead and then wheeling them through the prison yard, past the other prisoners, up over the bridge, and to the crematorium.

We followed behind one of the loaded wagons to get a better look at what was going on. There, alongside the building, several carts were lined up, waiting their turns to be unloaded. As each was maneuvered close to one of a series of windows, the bodies were pulled off, stripped, and then tossed into the rooms. Clothing and shoes were then loaded back onto the carts for disposal in another part of the camp.

I was standing near the building when two U.S. Army nurses approached. "I'd recommend you not look," I said as they strained to see what was inside. They ignored me and continued to move toward the windows. "I don't think you should look in there," I warned again.

"We can handle ourselves, soldier," one of them said.

"I'm telling you, you don't want to see this," I said.

They peered into the darkened room. As their eyes adjusted, horror registered on their faces. The staggered back from the window. One of them fainted. Two soldiers and a doctor, walking nearby, ran to assist. "I told them not to look," I said.

"Okay," the doctor said as he knelt and revived the nurse. He looked up at the other one, who was covering her eyes with her hands, still repulsed by what she had witnessed. "You don't need to be looking in there!" he said. "That's not going to do you any good. Get back and help with the living!"

Inside, corpses were stacked floor to ceiling like cordwood. The workers were methodically going about the business of cremating the bodies, trying to keep pace with the increasing numbers being tossed through the windows from the carts outside. One man would open the oven door, revealing the flames and intense heat within. Another would take a specialized instrument, something like a modified boiler-room poker with a half circle welded to one end, grab a dead prisoner by the crotch, lift him or her onto a metal stretcher, and shove the body into the firebox.

The other worker then shut the door just long enough for his partner to grab the next body and lift it from the stack. Within seconds he opened the heavy furnace door and pulled out an ash-covered stretcher. And the process began again. And again. And again. A dozen ovens were in operation side by side as smoke billowed from the chimneys of the crematorium and filled the air above the camp.

Unloading corpses at the crematorium in Dachau, Germany, April 29, 1945.

Body pile inside the crematorium at Dachau, Germany, April 29, 1945.

"What are they saying?" Duthie asked as we approached the fence. We had walked back around to where the prisoners were. Some of them were talking to a few American soldiers in Italian, and I wanted to investigate.

"Let's see if this makes sense in English," I said. "That one said that we are a healthy, young, beautiful army and that we should stay away from the prisoners because they have diseases. And this one talking right now—let me listen—yeah, he's saying that there's a building somewhere over here where the Germans would take them and torture them . . . do medical experiments and things to them."

"*Quello,*" one of them said, pointing to a row of buildings at the end of the yard.

"They want us to go and see if anybody is alive in there," one of the Americans said. "Okay." He turned and assured the prisoners, "*Andiamo a vedere.* We'll go take a look."

"You can speak Italian?" the GI asked me as we walked away.

"Oh, hell yeah," I explained. "I could speak Italian before I could speak English. Didn't know English until I went to school in the first grade."

"Same here," he said, extending his hand. "Gallo. New York."

"Sacco. Alabama."

"Alabama? They got wops there?"

"A few."

"What the hell is this?" he asked as we arrived.

Averitt pointed to a sign written in German above the door. "I think it's some type of infirmary." Two soldiers were running out, handkerchiefs covering their mouths.

Inside we found several squalid, filthy rooms, each containing gruesome evidence of atrocities almost unimaginable. Prisoners lay dead, tied to steel-framed tables, injected with pus and rotting meat and infection, cut open without anesthetics, internal organs hanging out and spilling onto the floor. Some were suspended from hooks, weights tied to their feet, scales measuring the maximum load the body could hold before rupturing. Human bones were crushed in vises that recorded the pounds of force necessary to break them. Stacks of human skin were labeled in German, but easy enough to translate—YOUNG BOYS, MEN, WOMEN UNDER 30, and TATTOOS. Some of the tattooed ones were stretched over wires and hung against a window as decorations. On a counter across the room, human heads, staring vacantly into space, floated in jars of formaldehyde.

"Let's get the hell outta here," one of the guys said. We backed out of the building as fast as we could. Several of the guys threw up as soon as we got outside.

Around on the other side of the complex, but still within view of the prisoners, was a relatively modern, well-equipped military hospital intended solely for the use of the Nazis and their families. American soldiers were in the building, going room to room, evacuating all the German patients at gunpoint, regardless of their condition, and sending them out into the prison yard below. The American GIs were then escorting sick and dying prisoners inside the facility, giving them beds and a comfortable place to recover—or, for some, a more respectable and humane place to die. Some German doctors refused to assist the dying Jewish prisoners and were arrested and led away by the U.S. military.

At the end of the row of buildings was a fenced kennel in which seventy-five or eighty vicious dogs, mostly Dobermans and German shepherds, paced, barked, and growled. Some of the prisoners had told us that the dogs were trained to kill the inmates by ripping the flesh from their bones and were released in packs to track down an escapee or, on occasion, just to run through the crowd and kill for sport. A small group of infantrymen approached the fence, aimed their weapons, and slaughtered the angry, defiant animals. The dogs snarled and howled maniacally as they died.

After several hours I found that I was walking around in a fog, unable to comprehend the magnitude of what I was witnessing in this place of unthinkable horrors. I found some steps and sat alone, gathering my thoughts. My mind and heart were racing. I wondered how Silverman or Chandler would have reacted to all this. I shut my eyes, took a few deep breaths, and tried to imagine I was somewhere else, but the visions of misery and the malodor of death were too intense to ignore.

I opened my eyes again and looked around. There, off to the right, I noticed a huge mound of shoes. Beside it was another, even larger, of striped clothing. As I lifted my gaze above the mounds, beyond the prisoners now quiet and resting in the yards, beyond their over-crowded and foul wooden barracks, up toward the sunset, I could see smoke still rising from the chimneys of the crematorium. This was a scene I could have never imagined back on the farm in Alabama, back in what seemed like another life.

At sunrise the next morning, the U.S. military marched the entire adult population of the town of Dachau through the concentration camp. Most of the townspeople were staring in disbelief at what they saw. Many covered their mouths with handkerchiefs or towels. Others wept openly at the atrocities laid out before them. Still others stared straight ahead and walked emotionlessly through the carnage.

"Wouldn't you just like to beat the shit out of every one of 'em?" Sarge asked as we watched them file past.

"I will if you will," Chicago responded.

"Me, too," I said. Spotted Bear and Sam Martin nodded in agreement.

Averitt looked up at the smoke still pouring from the chimneys and lingering over the town. "They had to know," he said.

We left Dachau later that morning, having assisted the relief organizations in their efforts to bring some semblance of humanity back to this place that had endured so much tragedy and sorrow. Our hearts were burdened by what we had seen but somehow lifted by the passionate gratitude of the liberated, who had blessed us with their joy and rewarded us with unexpected insights into what we had achieved.

German civilians being marched through the camp at Dachau, Germany, April 30, 1945.

Now, after a year of combat, each of us finally and forever understood why destiny had called us to travel so far from the land of our birth and fight for people we did not know. And so it was here, in this place abandoned by God and accursed by men, that we came to discover the meaning of our mission.

Ready to go home (I'm in the back). Salzburg, Austria, November 1945.

PART IV

THE LONG JOURNEY HOME

By God, I am proud of these men. To me, it's a never-ending marvel what our soldiers can do. Now, some people will say that sounds like I'm bragging about what a great man George Patton is, but I did not have anything to do with it. The people who actually did it are the soldiers of the Third Army. The soldier is the Army.

This war has made higher demands on courage and discipline than any war of which I have known. But when you see men who have demonstrated courage and discipline, and they are killed and wounded, it naturally raises a lump in your throat and sometimes produces a tear in your eye.

Back of us stretches a line of men whose acts of valor, of self-sacrifice and of service have been the theme of song and story since long before recorded history began. These are your fathers and grandfathers. These are the victors of war.

—GENERAL GEORGE S. PATTON

19

Victory

We had plenty of work as we rolled southward through Munich toward the Austrian border, doing our best to keep up with the infantry. We stopped briefly in a few towns, but just barely long enough to set up some phones before pulling out again. And we didn't have time to dally around, because our new driver, Chicago, made Tex seem like a patient man.

On the road to Salzburg, we encountered thousands of surrendering German troops. They filed past us in an endless procession, carrying white flags, looking almost happy that they were our prisoners and that the war was over for them. Word was passing through the ranks that the entire Nazi army had given up, with the exception of a few rogue SS troopers in the mountains. As darkness descended over the convoy, Captain English came by in a jeep and told us it was okay to turn on our headlights. Many of the drivers coupled that with blowing their horns.

We arrived in the beautiful city of Salzburg, Austria, on the morning of May 3, 1945, amid the welcoming cheers and open arms of its citizens. Some of the soldiers were walking in the streets, shaking hands with the men, hugging the older women, and kissing the young ones. Others were crowded atop the trucks and jeeps and tanks, holding their weapons aloft, and accepting the adulation of the people.

A minor traffic jam of men and vehicles forced us to a momentary stop at the city's main gate. An elderly man, standing on the base of a fountain and obviously a Nazi sympathizer, yelled, "I don't know how you Americans won the war, because you don't have any discipline!"

Several of the soldiers looked at him. Spotted Bear, standing on top of the truck cab, responded in no uncertain terms. "You can take all that Nazi discipline and shove it up your ass!"

The soldiers and civilians held their hands high in the air and cheered wildly as the convoy once again started moving. The people were clapping in unison as we passed under the archway into the city. The old man had disappeared from his perch and was nowhere to be seen.

We took over some houses beside a lake on the edge of Salzburg. Most of them were owned by Nazi officers, long since arrested or killed in action. Their families, still occupying the dwellings, were ordered out. One German officer remained in his home when the infantry arrived. As they approached the residence, the Americans heard four gunshots. When they entered, they found the officer and his family—wife, son, and daughter—all dead in the parlor.

We didn't stay in that place. We were assigned one about three houses down, a nice, big one with two stories, lots of rooms, and a couple of fireplaces. The Army brought in extra mattresses, pillows, sheets, and blankets so that we could each have our own bed. It felt good. Very good.

We spent the next day or two getting organized, helping set up a headquarters, and hooking up some switchboards for Carapella and his

**Our quarters in
Salzburg, Austria.**

men. But mostly we rested. Salzburg, set like a jewel among the Austrian Alps, afforded us a sense of confidence and relaxation, a decompression of sorts from the horrors of battle to a lightness of being in one of the most beautiful places on earth. The clean mountain air—and the lack of bullets flying through it—was refreshing and healing.

On the calm and peaceful night of May 5, in our new battalion headquarters on a hill overlooking the city, we gathered around a radio for an important speech from President Harry S. Truman. President Roosevelt had died in April, and this was the first time any of us had heard Truman's voice. Now we were listening intently as he announced to us and to the world that the war in Europe was officially over.

Captain English told us that General Haislip had issued orders that we were not to behave improperly or celebrate by firing our weapons after the announcement. We immediately ran into the streets, holding up our rifles, shooting them repeatedly while shouting at the top of our lungs. Bright tracer bullets lit up the night skies over Salzburg. Guys were jumping, running around, hugging guys from other units, hollering, then stepping back and shooting another few rounds in the air. It was amazing. It was the only way we knew to express how we felt.

Cardini came up and put his arms around me. "We're going home, Sacco!" he said. "We're going home!"

We *were* going home, but not right away. The war was over, but not the work. Having strung cable across Europe, we were now given the responsibility of establishing a dependable phone system that would link the Allied military in all sectors. The first order of business was to locate any existing telephone facilities and then dismantle and reconfigure them to suit our needs. Some guys were sent out to man telephone companies in the different towns and villages surrounding Salzburg. Sarge gave me the assignment of going to the former Nazi headquarters in the town square and disconnecting their lines. A couple of the guys with me were taking their time, carefully unscrewing each jack from the wall and untangling the cords for each phone.

"Here," I said. "This is gonna take forever if you do it that way. Let me show you how to speed up the process." I grabbed the wire and, with a quick jerk, yanked it from the wall. "Okay, where's the next one?"

We went down the line, pulling the wires out of the wall in each room and tossing the phones out the windows into the square. We then came through the building, installed American phones in their places, and spliced the main lines from the building to the switchboards on the hill.

In the meantime Duthie, Averitt, Hodges, and Hutter were given orders to go with some infantrymen up into the mountains and flush out a group of renegade SS troopers who had been sniping the city. "Orders are orders," Hodges lamented as he prepared to leave. "But I'm not that thrilled with the prospect of getting shot *after* the damn war!"

Sarge came around a few minutes later with the good news that the SS had walked into Salzburg and surrendered on their own. Hodges sat down behind a desk and smiled. "I think I'll have myself a beer . . . or two . . . or three . . . or four."

Grand Hotel Resort
Austrian Alps
May 7, 1945

The next morning a few of us were sent south of town to install telephones and lines for the 101st Airborne Division. When we arrived, the colonel informed us that we were to be billeted in the Grand Hotel, a luxury resort high in the Alps. The work was somewhat mundane, but the accommodations were outstanding—great food, clean rooms, comfortable beds, maid service, a spectacular view, and not that much to do.

We were a victorious army, barely out of our teens, with too much leisure time and an irrepressible urge to pull pranks on anyone who let down his guard. Such was the fate of a rather straitlaced lieutenant from Oklahoma named Prister. We probably would have done something to him based on his name alone, but in Prister's case he was always quoting the Bible to us, preaching about how we had no morals, and commenting that all the Jews and Catholics were going to hell.

Averitt came up with the plan. About ten of us, including several Airborne guys, chipped in to hire a prostitute and bring her up from

Salzburg. Prister's room was on the ground floor, so we hid in the bushes under his window as the girl knocked on his door.

"Come in," he said, thinking it was one of his men. The girl, who happened to be a knockout, opened the door and entered. We struggled not to laugh.

"What can I do for you?" he asked, peering over his reading glasses.

She didn't say a word. Instead she moved closer to him and began undressing, caressing each article of clothing across his face as she took it off. We were holding our hands to our mouths to keep from bursting into laughter. Occasionally he would look back down at his Bible, but the dangling of a stocking or the twirling of a blouse did get his attention. He took special notice of the way she spilled out of her bra. We all did.

Once she was completely nude, he looked her up and down, took off his glasses, closed the Good Book and said, "Young lady, are you done?"

She looked a bit confused. Pretty and naked, but confused.

"Are you finished?" he asked again.

She pursed her lips, nodded, then began picking up her clothes. We were outright laughing, running away from the window so as not to get caught. He had to know we were the culprits, although he never said a word about it.

"Let's go up to the Eagle's Nest," Duthie suggested one morning, referring to Hitler's mountain hideaway in nearby Berchtesgaden. "Some of the Airborne guys are going up in a few minutes to take a look around. Let's hurry and get a ride with them."

The Eagle's Nest had been bombed repeatedly by Allied planes. The remarkably solid structure was still intact, although leaning slightly to one side. As we wandered around within the dictator's residence, guys were laughing, going through his belongings, holding his clothes up in front of them while putting part of a comb to their mouth to imitate his mustache. Some were looking intently at the paintings and photographs on the walls, rummaging through his desk, or walking the grounds taking pictures.

I wandered into what must have been Hitler's study. A few infantrymen were poking around the room. A red flag with a black swastika hung in a corner. A large, boastful painting of the Führer hung beside it. An old typewriter sat on one end of a large, cluttered wooden desk. Books lined the opposite wall, but most of them were covered with dust and looked like they had never been read. A few on a table nearby were well worn, notes marked and scribbled throughout.

A large, overstuffed chair was positioned beside a picture window. I took a seat. From this vantage point, looking out at the commanding view, the world seemed a peaceful place. But, I kept thinking, it was from here, from this very room, that the demon himself must have gazed out over the mountains, fully aware that somewhere out there was Dachau, that somewhere out there was a place where innocent people were suffering and dying by his orders. Perhaps he sat here in this warm, comfortable chair and leisurely read a book, unmoved by the sorrow in the valleys below, untouched by what was happening beyond his pleasant and immediate view. Images of the agony and echoes of the screams flooded my mind. I stood, picked up a small globe from the table, and spun it around. I could feel myself hyperventilating with anger.

"There's Sacco," I heard Averitt's voice as he and Duthie entered the room. "Whoa, would you look at this shit! Hey, Sacco, don't you think the Führer would shit if he knew some good old American soldiers were in here going all through his stuff? Ain't that a hoot?"

"Yeah, that's a hoot all right," I said as I spun the globe again. "What an asshole. Let's get outta here."

"Yeah, let's go," Duthie said. "We need to be getting back."

"Too bad the son of a bitch shot himself," Averitt commented as we left, "seeing as how Patton wanted to be the one to put a bullet in his head."

Salzburg, Austria
May 27, 1945

The guys back in Salzburg had wasted no time in establishing a series of postwar conveniences and diversions. There was a post exchange, or PX, where we could buy anything we needed or wanted. Cigarettes were five cents per pack. Candy bars, soap, and toothpaste were three cents each. There was also a place called the Mozart

Red Ass Bar ticket
booklet, Salzburg,
Austria, May 1945.

Theater, where we could go to see newsreels of our victory and all the
Abbott and Costello movies we'd missed during the war. Perhaps most
important, they had set up a club downtown known as the Red Ass
Bar, where we could go, have a drink, hang out, and talk.

"When are you going back to see Monique?" Cardini asked.

"Captain said I could go in the next few weeks, soon as we finish
with that phone company in Linz," I answered. "Right, Sarge?"

"Yeah," he said. "I'll talk to Captain and make sure you're good to
go. You and Duthie going up to Linz, right?"

"And Hutter."

"Yeah, and Hutter." Sarge got up. "Anybody want another beer?"
Spotted Bear raised his hand. "Okay, one. What about you Cardini?"
Sarge asked. "Averitt? Okay, three, four. Okay, five. That's all I can carry."

"Thanks, Sarge," a couple of the guys said.

"One time," I began, "I was up on a pole—somewhere in France—
anyway, I was splicing some wires together, and my phone rang. It
shocked the hell out of me. I thought I was gonna fall off the damn
pole. So I answered it quick before it rang again, and it was Captain
English. I said, 'Damn, Captain, I've got these wires in my hands, and
you just shocked the shit out of me!' "

The other guys were laughing. "What did he say?" Averitt asked.

"He apologized and said that he was looking for Sarge. I said, 'Well
he ain't up here on this pole.' And he said he was sorry and that he
would try another number."

"Should've told him to try Duthie's number," cracked Chicago.

"Thanks a lot," said Duthie.

Sarge was returning with the beers.

"Hey, Sarge, what the hell's wrong with the phone lines in Germany?"
Cardini asked. "I've been trying to get through to my buddy with the
Third Army headquarters back in Heidelberg."

"I think the damn Germans went behind us and cut or ripped up a lot of our lines," Sarge answered as he put the drinks on the table. "And I don't just mean the army. I mean the civilians, too. But not to worry. New divisions are coming through and repairing most of it now."

"What the hell was the German army doing there after we kicked their ass?" Averitt asked.

"Hey, buddy," Chicago said, "here's the way it is. We'd kick their ass and then move on. Then another group of Nazis would come along and get their asses kicked by the next wave of our guys. What? You think they just gave up when they saw *us* flying past 'em?"

Sarge lifted his glass to Chicago. "For the first time in the war, Melvin's right about something. Congratulations."

"Thanks, Sarge," Chicago said, laughing and accepting the toast.

"Melvin?" Spotted Bear asked. "Your damn name's Melvin?"

"No wonder he goes by 'Chicago,'" remarked Cardini.

"Yeah," Chicago said. "You knew that. What about it?"

"Good thing he's not from Little Rock," said Averitt, who then looked at Spotted Bear. "Although that would probably be an acceptable name for one of your people."

"Go to hell, paleface," Spotted Bear responded as the rest of us, including Sam Martin, laughed.

"No thanks. Been in hell for the last year," Averitt said without missing a beat. "Didn't care for it."

Spotted Bear lifted his glass. "I'll drink to that."

We were walking through Salzburg the next morning when we saw a convoy of Russians entering the town. They didn't get nearly the welcome we did, although there were a few curious people watching as they rumbled past. One truck went up on the curb as it turned a corner and hit an elderly man who was standing there with his granddaughter. The young girl and the others nearby managed to jump out of the way. The truck careened along without so much as slowing down.

Two Russian soldiers in the back of the next truck laughed, jumped from their vehicle, ran over to the man, who was now in the street, pulled him onto the sidewalk beside his hysterical granddaughter, and then ran to catch back up with the convoy.

We ran over to the man, flagged down some MPs, and made sure he got to some American medics.

"Communist bastards," Chicago growled. "Patton doesn't trust them, and neither do I."

A few days later Chicago and Cardini got their revenge. We all knew that the Russians were willing to pay exorbitant prices for common items, such as Mickey Mouse watches and American cigarettes. What we didn't know was how gullible they were.

Cardini approached three of them and offered to sell them a U.S. Army jeep belonging to the 92nd, complete with a mounted .50-caliber machine gun. The Russians eyed the vehicle with interest.

Chicago, sitting in the jeep, kept saying, "No, it's not for sale," as the bids from the Communists increased.

"Let's sell the damn thing," Cardini said. "What the hell do we care?"

"I don't know," Chicago whined. "We could get in trouble." Finally the Russians produced a bag of cash from the back of a truck and offered thirty thousand dollars in American money. Chicago took the bag and tossed the keys to the happy new owners.

Having tucked the money safely away, Cardini and Chicago walked a few blocks and then waved down some American MPs. "What's the problem?" one of the MPs asked as they pulled over.

"What's the problem?" Cardini responded. "Three damn Russians just stole our goddamned jeep, that's what's the problem!"

The MPs promptly turned their vehicle around, located the confused Russians, confiscated the jeep, and returned it to Cardini and Chicago.

"Thanks, guys," Cardini said. "We owe you one."

"Damn Russians," one of the MPs said. "Think they own the damn world!"

20

Always

Duthie, Hutter, and I had completed our work in Linz and returned to Salzburg. I finished up some duties there, got my long-promised pass from Captain English, and boarded a troop train for France. Work was slowing down, and some of the guys had decided to do some traveling, figuring they might not have another chance to see Europe anytime soon. I was more concerned with seeing Monique than with seeing another bombed-out city.

It was mostly my fault that I hadn't heard from her in weeks— well, mine and the Army's. We had been constantly on the move since before Nuremberg with little or no time to sit and write letters. When I did write, I didn't want to tell her about Silverman and Chandler, because I didn't want her to know how dangerous it was or to worry that I might be next. I had sent her several letters from Austria but hadn't gotten any responses. A couple of the guys were having the same problem with the mail in Europe. A letter seemed to be able to go to New York and back much quicker than it could go to the next village.

I tried to patch through a few times, but the phones were somewhat erratic and usually monopolized by the officers anyway. Taking over Europe was apparently much easier than getting it back in working order. Carapella said I'd have better luck if I just got on a damn train and went there, which was precisely what I was now doing.

I hoped I hadn't made her angry by anything I'd written. I didn't think so, but I knew she'd gotten upset once when we were together because I was laughing at a crack one of the guys had made about how the Frenchmen were a bunch of sissies. I thought it was funny. She didn't. "My papa is a Frenchman!" she had protested.

"Yeah, I know," I said. "But he's not a sissy. Besides that, we *like* him."

It took some soothing, but she calmed down. I think she might have been just playing with me, but she seemed awfully serious at the time.

I felt a rush of happiness and excitement as the train pulled into Fénétrange. The sky was blue, the air was warm, the birds were singing, and for the first time in months I could hear the strains of "The Blue Danube" clearly in my mind. I picked up my bag, took a deep breath of the fresh pine scent in the breeze, and began walking through the town, wondering if Monique would spot me coming from a distance and be completely surprised, imagining how proud she would be when she noticed the new sergeant stripes on the arm of my jacket.

No one was home when I arrived. I checked my watch, wondering where the family might be. I decided to walk next door.

Little Bernard was playing near the barn. He ran to the house to tell his mother I was coming. Before I could reach the door, Madame Renard had come outside, an apron around her waist, a towel over her shoulder, flour on her hands. I smiled and waved. She stood motionless as I approached, then took the final few steps and embraced me. I could see Monsieur coming out of the house, smoking his pipe. He slowly took the pipe from his mouth and walked toward us.

"She is gone," Madame said, tears streaming down her face onto my shoulder. "She is gone."

Her husband reached out his hand, put it on my other shoulder, and bowed his head. The children stood at a respectful distance and watched in silence.

They took me to Monique's grave later that day. Her parents accompanied us. I was confused. It was difficult to comprehend what had happened. I felt too many emotions to understand any of them.

Through tears, her father told me the horror of how three German soldiers had come out of the forest and raped her and beat her and how some of the villagers had found her. She had lived for two days, he told me in a voice barely audible. The Americans caught the Germans and killed them on the spot. He embraced me and wept.

After a while they left me alone. I stayed there for hours, through the afternoon and setting sun, thinking about her, talking to her, trying to figure it out, wondering why this had happened, knowing it could not be true, listening to the wind, angry at God, angry at man, sifting through memories, feeling an ever-increasing pain in my heart, and knowing that it would last a lifetime.

The next day I sat in the park, on the bench where we liked to sit and talk and watch the comings and goings of others, where we had looked up into the trees and hinted of a life together. I couldn't smell the pine now, I couldn't feel the warmth of the air or hear the gentle conversations of people as they passed. My mind and body were numb, and my eyes ached from within. I felt waves of pain and guilt and anger flooding over me. Sometimes they would subside long enough for me to breathe, but then the spirit calming would recoil in the detestable reality that she was gone.

I left a letter on her headstone. I knelt and kissed the small photograph beside her name, stood, and walked back through the town to the train station.

My Dear Monique,

I am so lost and alone as I write these words. I can't tell you how I thirst to say them to you in person, to have you here beside me now so that I can look once more into your eyes, to hold you so close that I can feel the beating of your heart touching mine, and to whisper so softly that you can hear my thoughts.

It would be impossible to describe all that has happened since the last time we saw each other. But I want you to know that I've thought of you and missed you every step of the way. There is so much left to say and do—so much that is now replaced by emptiness.

I don't understand why I lived through it all if not to be with you. How can I survive the pain of knowing that you are gone and that I was not here to protect you? How can I forgive myself? You were the light that dispelled the darkness, that kept my hope—and my heart—alive through it all. How can I go on when I think that this war, which brought us together, now keeps us forever apart?

More than anything, I want to hold you again. If I could embrace you just one more time, I would be happy. Just one more time.

Last night, as I sat on our bench in the park, I looked up into the sky as it sparkled above me. And somehow, through the stillness of the night, a silent voice seemed to whisper from beyond those tiny, shimmering windows of light. It was calming, consoling, reassuring me that you are okay—that you are home.

And so now I pray to God above that from His dwelling place beyond those stars and within our hearts, He will one day reunite us in eternity. And at that moment, our destinies and our worlds will again coincide. Then, there will again be laughter and joy and dance and the sweetest of song. Then, you will look into my eyes and I will look into yours. And I will watch as that beautiful smile silently forms on your lips.

Remember me gently, Monique. Until that day in heaven comes, remember me gently. Remember me making you laugh, taking your hand, touching your face, calling your name. Remember me stumbling to waltz, but learning the dance to be near you. Remember me looking into your eyes and wanting for you only peace and goodness and happiness. When you think of me, remember me gently speaking to that most quiet place in your heart, me telling you through time and space and war and unending heartache that I loved you then, I love you now, I love you always.

21

Many Battles Ago

Salzburg, Austria
July–November, 1945

My remaining few months in Salzburg became a confusion of sights and sounds and soldiers. As time went along, I found it easier to breathe and to relax and, eventually, to smile. Every one of the guys was understanding and helpful. We had all been through a lot, and we had all suffered great losses, so each of us knew how it felt. Often I would see a girl walking down the street or sipping coffee at a café or chatting with a friend, and for one split second—for one brief, devastating moment—I would think it was Monique. Sometimes I would find myself walking faster to catch up with her, just to make sure, just to see. Sometimes I would go up to the castle on the hill above town and sit there and look out over the valley. And I would trace the river as it wound its way through the city and beyond. And I would look up into the mountains. And wonder.

There wasn't much work to do. Most of the tough assignments had been completed within the first weeks of our arrival. As we prepared to ship out, fresh replacements were being sent from the States to man our positions. They had it made. Their jobs were pretty easy compared to the ones we had done. They didn't even have to carry a gun. Averitt made the comment that it was as if these new boys had enlisted for a European vacation. But none of us really minded that much. We were just grateful to be alive, though the price of living had been high.

They were sending the married soldiers home first. That included Duthie, who had befriended a two-year-old orphan girl named

**Joe Hutter and me in Salzburg,
Austria, summer of 1945.**

Ingrid. They said that her mother had been killed in a bombing raid
and her father had been killed by the Russians. Duthie somehow
managed to get hold of a parachute and made her some clothes. He
asked us to look after Ingrid, and he promised to send her a doll
from America.

On the evening of November 21, 1945, the rest of us convened in
the Red Ass Bar for one final night. We were swapping stories, making fun of each other, and talking about what we were going to
accomplish when we got back home. As we were talking, I began to
realize that these boys were my brothers. I was the youngest of the
group, and, whether they realized it or not, all of them had looked out
for me as we went through the war. I guess, in reality, we had looked
out for each other. We were brothers, and we had been through hell
together. And as I thought of leaving them, tears began to well up in
my eyes.

Sarge, who was sitting next to me, put his hand on my shoulder.
"It's okay, Joe," he said. The other guys patted me on the back. "It's
okay," they said. "We'll see each other again someday."

November 22, 1945

Little Ingrid was at the train station in Salzburg, waving from the arms of an Army nurse, wearing her parachute dress, clutching a doll we'd bought her at a shop in town. Other friends we'd made, both military and civilian, came to see us off, to say good-bye and thank you.

The train took us through Austria and the northern part of Italy, then into the port of Marseilles, on the southern coast of France, where we went through a few days of medical exams before boarding the ship that would take us home. The motion of the boat moving away from the dock—and the knowledge that the war was behind us—was exhilarating.

We sailed through the Mediterranean, past the Rock of Gibraltar, and out into the Atlantic. The crossing was uneventful until the night before we were to arrive in Boston, when the swelling seas erupted into a strong and perilous storm. Gale-force winds lashed the boat, and freezing waves crashed into its side, causing it to jolt, shake, and shudder as it tossed about. Belowdecks we held tightly on to our bunks to keep from being thrown across the floor. Many times the ship sounded like it was going to break into pieces. I grasped my cross and prayed to God that, after everything we'd been through, He would allow us to arrive safely home.

In the early-morning hours of December 13, the storm abated, the seas calmed, and the ship limped into port. I had my duffel bag packed and ready. I wanted to get abovedecks as soon as possible so that I could see America.

Final farewells at the Salzburg train station in November 1945.

Many residents of Boston—people we didn't even know—had braved the weather to come out and welcome us home. They were standing all along the dock, clapping, cheering, and waving flags. We navigated the ice-coated decks, made our way down the gangplanks, and set foot in the U.S.A. Many of the guys got down and kissed the ground. Men shook our hands. Women hugged and kissed us. Children gathered around and touched our coats. We felt like heroes.

Army personnel met us and escorted us to a large dining facility, where we were offered all the steaks we could eat. Most of the guys ate at least a couple. I ate two and drank a gallon of milk. Spotted Bear claimed to have eaten four. We told him that Chandler could have downed six without a problem.

When we'd had our fill, a civilian came into the room and told us there were buses waiting to take us to a building where a special phone bank had been set up so that we could call home. The operators there asked us where we wanted to call, then gave us a number, and told us to wait. We had been in the Army for three years, so we were very well trained in the art of waiting. When they finally called my number, I picked up the phone in front of me and heard it ringing on the other end.

Mama answered. It was an incredible feeling to hear her voice. "Joe!" she cried out. "Joe, is that you? Is it you?" She put Papa on the line, then Grandma Sacco. I was speaking in Italian, and some of the operators must have been able to understand what I was saying, because I saw them smiling. I talked a long time, even though they'd told us to be brief. I couldn't help it. I was home.

December 17, 1945

They put me in charge of eight other men on the train, giving me the responsibility of checking their papers and making sure they got off at the correct stops between Boston and Atlanta. They seemed like decent guys, but I didn't know any of them, so I sat by myself as the train made its way south.

In the solitude of those hours, as I watched the passing landscape, my mind went back to images of the war, of friendships forged and battles fought, of captives liberated and brothers left behind. I could see it all as if it had been a vivid dream, now replaying itself in my head. I could see Monique, and I could hear her whispering my name ever so softly, ever so sweetly. She smiled, reached out, and touched my hand.

Fort McPherson, Georgia
December 18, 1945

"Hot damn! Normandy. Omaha Beach. The Bulge. Rhineland. You were all over the European theater, weren't you, Sergeant Sacco?" the major asked as he processed my papers. "Ever run across Patton?"

"Oh, hell yeah," I answered. "Used to see him every day."

"Damn," he repeated. "Y'all did one helluva job over there, soldier."

"Thank you, sir." I was wondering where he had been during the fighting when he noticed something on my papers and spoke again.

"You . . . uh . . . Sergeant, you realize you had an exemption, don't you?"

"Huh?"

"It says here you're an only child and . . . yeah . . . and a farmhand."

"Yes, sir, I was," I said. "So?"

"Army gave exemptions for only children and farmhands. You didn't know that?"

"Nope."

"Maybe I shouldn't have told you," he said, straightening up the papers. "What the hell? It doesn't matter now, does it?"

I was surprised, but I laughed. "No, sir, I don't guess so."

"Let's get you outta here so you can catch your train," he said. "Damn fine work, soldier. Thank you."

The major walked me to the station and helped me carry my gear. "It's the least I can do," he had said. As we sat and waited on the train, he talked to me about what to expect when I got home, how things might be different, how people might not be able to understand what we had been through, and how I might have dreams that would wake me up at night. He said, "Don't forget that you're just twenty-one years old, son. You're still a young man, and your whole life's ahead of you. You've been through a lot. All the guys have. But things will smooth out."

As I was climbing aboard the train, he said, "Sergeant." I turned around, and he saluted me. I returned his salute. "Welcome home," he said.

A little boy, sitting across from me with his mother, kept staring at me. Finally he spoke. "Are you a soldier?"

"Yes."

"Were you in the war?"

"Yeah. I sure was."

"Did you kill any Germans?"

His mother chided him for the question, looked at me, and smiled. "Tell him we're proud of him and all the soldiers," she instructed.

"We're proud of you and all the other soldiers," he dutifully repeated.

"Thank you."

"Do you have a gun?"

"Nah, not anymore," I answered. "Not anymore."

As the train pulled closer to Birmingham, my heart started pounding and a very unexpected—but very real—fear came over me. Perhaps I wasn't the same person who had left three years earlier to join the Army. Perhaps I'd changed too much. Perhaps I would never be able to adequately express what had happened.

I could see Papa and Mama, Uncle Vincent, Uncle Joe, and some of the cousins standing on the platform, dressed in their Sunday best, straining to look into each railcar as we approached. I waited until the train came to a full stop. Other people were already up, getting their bags, and going. I helped the lady and her son retrieve their belongings, then put on my coat and collected my duffel bag from the rack.

I pulled out the silver crucifix Grandpa Sacco had given me before I left home. It had seen me through all the dangers of the war. Now I pressed it to my chest once again and prayed that the heartaches would heal.

My cousin Josephine noticed me first. "There he is!" she said, running toward me as I leaned out the door into the brisk December air. "There's Joe!"

The others were coming in my direction.

I stepped off the train, set my bag on the ground, and embraced the world I'd left behind so many battles ago.

Acknowledgments

Though solitary work it may be, the writing of a book does not occur in a vacuum. It takes the cooperation, encouragement, dedication, and inspiration of many others in addition to the author. Every book, therefore, not only tells a story, it has a story of its own. And so it has been with this one, beginning with a dinner party at which a friend noticed an Army photograph of my father taken in Salzburg, Austria, at the end of World War II.

A series of questions about his experiences followed, prompting me to pull out the pictures he and his buddies had taken at Dachau.

"You've got to write a book about this," one of my friends, Vivian Stubbers, enthusiastically suggested.

"Nah," I answered in my wisdom.

Others in attendance concurred with Vivian's idea.

Early the next morning, I contacted another close friend, Cal Thomas, and asked his opinion. From that point on, Cal became a guide and mentor, constantly encouraging me, challenging me, teaching me, lifting my spirits, and helping me make the dream a reality. My debt to Cal cannot be adequately stated, let alone repaid.

Without his professionalism and strength of character to light the way, without his constant encouragement, sense of humor, and positive outlook even in the darkest of times, this book would have never been written. Cal and his wife, Ray, were with me every step of the way, offering their support, friendship, counsel, and even a place to rest my weary head when visiting Washington, D.C.

As I thank them both, I must in turn thank the wonderful friend who introduced us—the legendary actress and eternal star Loretta Young. From the first day of my arrival in Los Angeles, Loretta assumed the role of my "West Coast mom." She encouraged me, prayed for me, cooked for me, introduced me to her friends, and spent countless hours talking with me, telling me stories of her life and the Golden Age of Hollywood. Most of all she believed in me. There was never a truer friend than Loretta Young. She is an important element in the writing of this book, not only because of her introduction to Cal and Ray Thomas, but also because of her unwavering faith in me and my goals. She will be missed by all of us who knew her well and loved her.

This book could not have been written without the gracious help and support of the men whose story comes to life within its pages—the men of the 92nd Signal Battalion. Jim Hodges, Ed Duthie, Paul Averitt, Mike Carapella, Marty Kesselman, Mark Fox, Wilbur Fulton, Al Spagnoletti, Dom Lipari, Owen Wallace, Robert Jud, Bill Miller, Ace Azar, Lloyd Mallory, and Virgil Bredow, along with the families of Joe Hutter and (First Sergeant) Ernest Thomas, contributed memories, photographs, and accounts of the war. Their only request was that I write as honestly as I could about their journey, their daily struggle for survival, and the transformation that took place within them as their mission unfolded. These men were ordinary soldiers who, through the course of the war, became ordinary heroes—the type we take for granted these days.

Jim Hodges was the point man in this effort, collecting usable information from all the guys and then relating it—along with his own experiences—to me in such an organized and coherent fashion that I was easily able to see how all the pieces fit together. Without Jim's energetic assistance, I might still be sifting through hundreds of slightly different versions of the story, trying to figure out what really happened.

I was fortunate enough to have been a special guest of the 92nd Signal Battalion at their annual reunion in August 2001. It was there that I had the pleasure of personally meeting many of the book's characters. Over the course of the reunion, I also had the opportunity to carefully observe the dynamics within the group, as well as individual personalities and even speech patterns. It was an invaluable experience in the construction and ultimate feel of this work.

Quite obviously, the person to whom I owe the greatest thanks is my father, Joe Sacco. Without his stories and inspiration, there would be no book, and without his ability to survive the perils of a world war, there would be no me. My mother, Rosalie Sacco, helped make this work a reality with her insights, constant encouragement, prayers, and baked goods. My sister Rosa-Marie Sacco assisted by doing a great deal of early research and by collecting essential information from the U.S. Army that helped me trace the journey of the men through the war. My brother, John Sacco; sisters Lucy Barry and Gina Wittig; and brother-in-law Bill Wittig each offered unselfish support precisely when it was most needed along the way.

None of this would have been possible without Judith Regan, who had faith in me when others didn't, and Cassie Jones, who nurtured the book through its own journey from conception to completion. I have often heard authors effuse about the brilliance of their editors, but I must admit that I never quite understood what all the fuss was about. Now I know. Brian Saliba was chosen as my editor at ReganBooks, and I couldn't have wished for one better. This book would not have been the same without his patience, ideas, observations, and enthusiasm. His dedication and talent show through on each page. I am fortunate to have worked with him and even more fortunate to call him a friend.

I especially thank Rev. Matthew Brennan for his honest and uplifting spiritual direction—not necessarily in relation to the book itself but, more important, in relation to life. The courage to act is always preceded by the light of faith.

Throughout my life I have been blessed with good and loyal friends, not simply the fair-weather type. Each has played an important role in the realization of this book, offering perfectly timed encouragement and support every step of the way. My friends know who they are. And to each I am grateful. They stand with me in victory because they stood with me in times less certain.